OLIVE THORNE MILLER

A BIRD-LOVER
IN THE
WEST

Idle Winter Press
Portland, Oregon

Interior text content by Harriet Mann Miller (Olive Thorne Miller) was originally published in 1894
by Houghton, Mifflin & Co., The Riverside Press, Cambridge Massachusetts, U.S.A.;
electrotyped and printed by H. O. Houghton & Co. and is in the public domain.
Bird species common and scientific names have been modernized where applicable.

Cover background photograph is in the public domain.

Idle Winter Press
Portland, Oregon
http://IdleWinter.com

First published 1894
This edition published 2015
Printed in the United States of America

The text of this book is in Alegreya

ISBN-13: 978-0692371916 (Idle Winter Press)
ISBN-10: 0692371915

INTRODUCTORY

The studies in this volume were all made, as the title indicates, in the West; part of them in Colorado (1891), in Utah (1893), and the remainder (1892) in what I have called "The Middle Country," being Southern Ohio, and West only relatively to New England and New York, where most of my studies have been made.

Several chapters have appeared in the "Atlantic Monthly" and other magazines, and in the "Independent" and "Harper's Bazar," while others are now for the first time published.

OLIVE THORNE MILLER

CONTENTS

IN THE ROCKY MOUNTAINS

IN THE MIDDLE COUNTRY

IN THE ROCKY MOUNTAINS

Trust me, 't is something to be cast
Face to face with one's self at last,
To be taken out of the fuss and strife,
The endless clatter of plate and knife,
The bore of books, and the bores of the street,
From the singular mess we agree to call Life.
And to be set down on one's own two feet
So nigh to the great warm heart of God,
You almost seem to feel it beat
Down from the sunshine and up from the sod;
To be compelled, as it were, to notice
All the beautiful changes and chances
Through which the landscape flits and glances,
And to see how the face of common day
Is written all over with tender histories.

James Russell Lowell

I. CAMPING IN COLORADO

This chronicle of happy summer days with the birds and the flowers, at the foot of the Rocky Mountains, begins in the month of May, in the year eighteen hundred and ninety-two.

As my train rolled quietly out of Jersey City late at night, I uttered a sigh of gratitude that I was really off; that at last I could rest. Up to the final moment I had been hurried and worried, but the instant I was alone, with my "section" to myself, I "took myself in hand," as is my custom.

At the risk of seeming to stray very far from my subject, I want at this point to say something about rest, the greatly desired state that all busy workers are seeking, with such varying success.

A re-creative recreation I sought for years, and

> "I've found some wisdom in my quest
> That's richly worth retailing,"

and that cannot be too often repeated, or too urgently insisted upon. What is imperatively needed, the sole and simple secret of rest, is this: To go to our blessed mother Nature, and to go with the whole being, mind and heart as well as body. To deposit one's physical frame in the most secret and sacred "garden of delights," and at the same time allow the mind to be filled, and the thoughts to be occupied, with the concerns of the world we live in year after year, is utterly useless; for it is not the external, but the internal man that needs recreation; it is not the body, but the spirit that demands refreshment and relief from the wearing cares of our high-pressure lives. "It is of no use," says a thoughtful writer, "to carry my body to the woods, unless I get there myself."

Let us consult the poets, our inspired teachers, on this subject. Says Lowell,—

"In June 't is good to lie beneath a tree
While the blithe season comforts every sense,
Steeps all the brain in rest, and heals the heart,
Brimming it o'er with sweetness unawares,
Fragrant and silent as that rosy snow
Wherewith the pitying apple-tree fills up
And tenderly lines some last-year's robin's nest."

And our wise Emerson, in his strong and wholesome, if sometimes rugged way,—

"Quit thy friends as the dead in doom,
And build to them a final tomb.
Behind thee leave thy merchandise,
Thy churches and thy charities.

> Enough for thee the primal mind
> That flows in streams—that breathes in wind."

Even the gentle Wordsworth, too; read his exquisite sonnet, beginning,—

> "The world is too much with us; late and soon,
> Getting and spending, we lay waste our powers."

All recognize that it is a mental and spiritual change that is needed.

With the earnest desire of suggesting to tired souls a practicable way of resting, I will even give a bit of personal history; I will tell the way in which I have learned to find recreation in nature.

When I turn my back upon my home, I make a serious and determined effort to leave behind me all cares and worries. As my train, on that beautiful May evening, passed beyond the brick and stone walls, and sped into the open country, and I found myself alone with night, I shook off, as well as I was able, all my affairs, all my interests, all my responsibilities, leaving them in that busy city behind me, where a few burdens more or less would not matter to anybody. With my trunks checked, and my face turned toward the far-off Rocky Mountains, I left the whole work-a-day world behind me, departing—so far as possible—a liberated soul, with no duties excepting to rejoice and to recruit. This is not an easy thing to do; it is like tearing apart one's very life; but it can be done by earnest endeavor, it has been done, and it is a charm more potent than magic to bring restoration and recreation to the brain and nerve-weary worker.

To insure any measure of success I always go alone; one familiar face would make the effort of no avail; and I seek a place where I am a stranger, so that my ordinary life cannot be recalled to me. When I reach my temporary home I forget, or at least ignore, my notions as to what I shall eat or drink, or how I shall sleep. I take the goods the gods provide, and adjust myself to them. Even these little things help one out of his old ways of thought and life. To still further banish home concerns, I mark upon my calendar one week before the day I shall start for home, and sternly resolve that not until I reach that day will I give one thought to my return, but will live as though I meant to stay always. I take no work of any sort, and I banish books, excepting a few poets and studies of nature.

Such is the aim of my honest and earnest striving; that I do not quite reach my goal is merely to say I am human. Letters from home and friends will drag me back to old interests, and times will come, in sleepless nights and unguarded moments, when the whole world of old burdens and cares sweep in and overwhelm me. But I rouse my will, and resolutely, with all my power, push them back, refuse to entertain them for a moment.

The result, even under these limitations, is eminently satisfactory. Holding myself in this attitude of mind, I secure a change almost as complete as if I stepped out of my body and left it resting, while I refreshed myself at the fountain of life. A few weeks in the country make me a new being; all my thoughts are turned into fresh channels; the old ruts are smoothed over, if not obliterated; nerves on the strain all the year have a chance to recreate themselves; old worries often weaken and fade away.

The morning after I left home that balmy evening in May dawned upon me somewhere in western New York, and that beautiful day was passed in speeding through the country, and steadily getting farther and farther from work and care.

And so I went on, day after day, night after night, till I entered Kansas, which was new to me. By that time I had succeeded in banishing to the farthest corner of my memory, behind closed and locked doors, all the anxieties, all the perplexities and problems, all the concerns, in fact, of my home life. I was like a newly created soul, fresh and eager to see and enjoy everything. I refused the morning papers; I wished to forget the world of strife and crime, and to get so into harmony with the trees and flowers, the brooks and the breezes, that I would realize myself

> "Kith and kin to every wild-born thing
> that thrills and blows."

In one word, I wished as nearly as possible to walk abroad out of my hindering body of clay.

I looked out of the windows to see what the Cyclone State had to give me. It offered flowers and singing birds, broad fields of growing grain, and acres of rich black soil newly turned up to the sun. Everything was fresh and perfect, as if just from the hands of its maker; it seemed the paradise of the farmer.

From the fertile fields and miles of flowers the train passed to bare, blossomless earth; from rich soil to rocks; from Kansas to Colorado. That part of the State which appeared in the morning looked like a vast body of hardly dry mud, with nothing worth mentioning growing upon it. Each little gutter had worn for itself a deep channel with precipitous sides, and here and there a great section had sunken, as though there was no solid foundation. Soon, however, the land showed inclination to draw itself up into hills, tiny ones with sharp peaks, as though preparing for mountains. Before long they retreated to a distance and grew bigger, and at last, far off, appeared the mountains, overtopping all one great white peak, the

"Giver of gold, king of eternal hills."

A welcome awaited me in the summer home of a friend at Colorado Springs, in the presence of the great Cheyenne Range, with the snow-cap of Pike's Peak ever before me. Four delightful days I gave to friendship, and then I sought and found a perfect nook for rest and study, in a cottonwood grove on the cool banks of the Minnelowan (or Shining Water). This is a mad Colorado stream which is formed by the junction of the North and South Cheyenne Cañon brooks, and comes tumbling down from the Cheyenne, rushing and roaring as if it had the business of the world on its shoulders, and must do it man-fashion, with confusion and noise enough to drown all other sounds.

Imagine a pretty, one-story cottage, set down in a grove of cottonwood-trees, with a gnarly oak and a tall pine here and there, to give it character, and surrounded as a hen by her chickens, by tents, six or eight in every conceivable position, and at every possible angle except a right angle. Add to this picture the sweet voices of birds, and the music of water rushing and hurrying over the stones; let your glance take in on one side the grand outlines of Cheyenne Mountain,

"Made doubly sacred by the poet's pen
And poet's grave,"

and on the other the rest of the range, overlooked by Pike's Peak, fourteen thousand feet higher than the streets of New York. Do this, and you will come as near to realizing Camp Harding as one can who is hundreds of miles away and has never seen a Colorado camp.

Do not think, however, that such camps are common, even in that land of outdoors, where tents are open for business in the streets of the towns, and where every householder sets up his own canvas in his yard, for the invalids to sleep in, from June to November. The little settlement of tents was an evolution, the gradual growth of the tent idea in the mind of one comfort-loving woman. She went there seven or eight years before, bought a grove under the shadow of Cheyenne, put up a tent, and passed her first summer thus. The next year, and several years thereafter, she gradually improved her transient abode in many ways that her womanly taste suggested,—as a wooden floor, a high base-board, partitions of muslin or cretonne, door and windows of wire gauze. The original dwelling thus step by step grew to a framed and rough-plastered house, with doors and windows en règle.

Grouped picturesquely around the house, however, were some of the most unique abiding-places in Colorado. On the outside they were permanent tents with wooden foundations; inside they were models of comfort, with regular beds and furniture, rugs on the floor, gauzy window curtains, drapery wardrobes, and even tiny stoves for cool mornings and evenings. They combined the comforts of a house with the open air and delightful freshness of a tent, where one might hear every bird twitter, and see the dancing leaf shadows in the moonlight. Over the front platform the canvas cover extended to form an awning, and a wire-gauze door, in addition to one of wood, made them airy or snug as the weather demanded.

The restfulness craved by the weary worker was there to be had for both soul and body, if one chose to take it. One might swing in a hammock all day, and be happy watching "the clouds that cruise the sultry sky"—a sky so blue one never tires of it; or beside the brook he might "lie upon its banks, and dream himself away to some enchanted ground." Or he might study the ever-changing aspect of the mountains,—their dreamy, veiled appearance, with the morning sun full upon them; their deep violet blueness in the evening, with the sun behind them, and the mystery of the moonlight, which "sets them far off in a world of their own," as tender and unreal as mountains in a dream.

He might do all these things, but he is far more likely to become excited, and finally bewitched by guide-books, and photographs, and talk all about him of this or that cañon, this or that pass, the Garden of the Gods, Manitou, the Seven Sisters' Falls, the grave of "H. H.;" and unless a fool or a philosopher, before he knows it to be in the full swing of sight-seeing, and becoming learned in the ways of burros, the "Ship of the Rockies," so indispensable, and so common that even the babies take to them.

This traveler will climb peaks, and drive over nerve-shaking roads, a steep wall on one side and a frightful precipice on the other; he will toil up hundreds of steps, and go quaking down into mines; he will look, and admire, and tremble, till sentiment is worn to threads, purse depleted, and body and mind alike a wreck. For this sort of a traveler there is no rest in Colorado; there always remains another mountain to thrill him, another cañon to rhapsodize over; to one who is greedy of "sights," the tameness of Harlem, or the mud flats of Canarsie, will afford more rest.

For myself I can always bear to be near sights without seeing them. I believed what I heard—never were such grand mountains! never such soul-stirring views! never such hairbreadth roads! I believed—and stayed in my cottonwood grove content. I knew how it all looked; did I not peer down into one cañon, holding my breath the while? and, with slightly differing arrangement of rocks and pine-trees and brooks, are not all cañons the same? Did I not gaze with awe at the "trail to the grave of H. H.," and watch, without envy, the sight-seeing tourist struggle with its difficulties? Could I not supply myself with photographs, and guide-books, and poems, and "H. H.'s" glowing words, and picture the whole scene? I could, I did, and to me Colorado was a delightful place of rest, with mountain air that it was a luxury to breathe (after the machinery adjusted itself to the altitude), with glorious sunshine every morning, with unequaled nights of coolness, and a new flower or two for every day of the month.

If to "see Colorado" one must ascend every peak, toil through every cañon, cast the eyes on every waterfall, shudder over each precipice, wonder at each eccentric rock, drink from every spring, then I have not seen America's Wonderland. But if to steep my spirit in the beauty of its mountains so that they shall henceforth be a part of me; to inhale its enchanting air till my body itself seemed to have wings; if to paint in my memory its gorgeous procession of flowers, its broad mesa crowned with the royal blossoms of the yucca, its cosy cottonwood groves, its brooks rushing between banks of tangled greenery; if this is to "see Colorado," then no one has ever seen it more thoroughly.

The "symphony in yellow and red," which "H. H." calls this wonderland, grows upon the sojourner in some mysterious way, till by the time he has seen the waxing and waning of one moon he is an enthusiast. It is charming alike to the sight-seer whose jaded faculties pine for new and thrilling emotions, to the weary in brain and body who longs only for peace and rest, and to the invalid whose every breath is a pain at home. To the lover of flowers it is an exhaustless panorama of beauty and fragrance, well worth crossing the continent to enjoy; to the mountain lover it offers endless attractions.

Nothing is more fascinating to the stranger in Colorado than the formation of its cañons, not only the grand ones running up into the heart of the mountains, but the lesser ones cutting into the high table-land, or mesa, at the foot of the hills. The above mentioned cottonwood grove, for example, with its dozen of dwellings and a natural park of a good many acres above it, with tall pines that bear the marks of age, is so curiously hidden that one may come almost upon it without seeing it. It is reached from Colorado Springs by an electric road which runs along the mesa south of the town. As the car nears the end of the line, one begins to look around for the grove. Not a tree is in sight; right and left as far as can be seen stretches the treeless plain to the foot of the eternal hills; not even the top of a tall pine thrusts itself above the dead level. Before you is Cheyenne—grim, glorious, but impenetrable. The conductor stops. "This is your place," he says. You see no place; you think he must be mistaken.

"But where is Camp Harding?" you ask. He points to an obscure path—"trail" he calls it—which seems to throw itself over an edge. You approach that point, and there, to your wonder and your surprise, at your feet nestles the loveliest of smiling cañon-like valleys, filled with trees, aspen, oak, and pine, with here and there a tent or red roof gleaming through the green, and a noisy brook hurrying on its way downhill. By a steep scramble you reach the lower level, birds singing, flowers tempting on every side, and the picturesque, narrow trail leading you on, around the ledge of rock, over the rustic bridge, till you reach the back entrance of the camp. Before it, up the narrow valley, winds a road, the carriage-way to the Cheyenne cañons.

II. IN THE COTTONWOODS

A cottonwood grove is the nearest approach to our Eastern rural districts to be found in Colorado, and a cotton storm, looking exactly like a snowstorm, is a common sight in these groves. The white, fluffy material grows in long bunches, loosely attached to stems, and the fibre is very short. At the lightest breeze that stirs the branches, tiny bits of it take to flight, and one tree will shed cotton for weeks. It clings to one's garments; it gets into the houses, and sticks to the carpets, often showing a trail of white footprints where a person has come in; it clogs the wire-gauze screens till they keep out the air as well as the flies; it fills the noses and the eyes of men and beasts. But its most curious effect is on the plants and flowers, to which it adheres, being a little gummy. Some flowers look as if they were encased in ice, and others seem wrapped in the gauziest of veils, which, flimsy as it looks, cannot be completely cleared from the leaves.

It covers the ground like snow, and strangely enough it looks in June, but it does not, like snow, melt, even under the warm summer sunshine. It must be swept from garden and walks, and carted away. A heavy rain clears the air and subdues it for a time, but the sun soon dries the bunches still on the trees, and the cotton storm is again in full blast. This annoyance lasts through June and a part of July, fully six weeks, and then the stems themselves drop to, the ground, still holding enough cotton to keep up the storm for days. After this, the first rainfall ends the trouble for that season.

In the midst of the cottonwoods, in beautiful Camp Harding, I spent the June that followed the journey described in the last chapter,—

"Dreaming sweet, idle dreams of having strayed
To Arcady with all its golden lore."

The birds, of course, were my first concern. Ask of almost any resident not an ornithologist if there are birds in Colorado, and he will shake his head.

"Not many, I think," he will probably say. "Camp birds and magpies. Oh yes, and larks. I think that's about all." This opinion, oft repeated, did not settle the matter in my mind, for I long ago discovered that none are so ignorant of the birds and flowers of a neighborhood as most of the people who live among them. I sought out my post, and I looked for myself.

There are birds in the State, plenty of them, but they are not on exhibition like the mountains and their wonders. No driver knows the way to their haunts, and no guide-book points them out. Even a bird student may travel a day's journey, and not encounter so many as one shall see in a small orchard in New England. He may rise with the dawn, and hear nothing like the glorious morning chorus that stirs one in the Atlantic States. He may search the trees and shrubberies for long June days, and not find so many nests as will cluster about one cottage at home.

Yet the birds are here, but they are shy, and they possess the true Colorado spirit,—they are mountain-worshipers. As the time approaches when each bird leaves society and retires for a season to the bosom of its own family, many of the feathered residents of the State bethink them of their inaccessible cañons. The saucy jay abandons the settlements where he has been so familiar as to dispute with the dogs for their food, and sets up his homestead in a tall pine-tree on a slope which to look at is to grow dizzy; the magpie, boldest of birds, steals away to some secure retreat; the meadow-lark makes her nest in the monotonous mesa, where it is as well hidden as a bobolink's nest in a New England meadow.

The difficulties in the way of studying Colorado birds are several, aside from their excessive suspicion of every human being. In the first place, observations must be made before ten o'clock, for at that hour every day a lively breeze, which often amounts to a gale, springs up, and sets the cottonwood and aspen leaves in a flutter that hides the movements of any bird. Then, all through the most interesting month of June the cottonwood-trees are shedding their cotton, and to a person on the watch for slight stirrings among the leaves the falling cotton is a constant distraction. The butterflies, too, wandering about in their aimless way, are all the time deceiving the bird student, and drawing attention from the bird he is watching.

On the other hand, one of the maddening pests of bird study at the East is here almost unknown,—the mosquito. Until the third week in June I saw but one. That one was in the habit of lying in wait for me when I went to a piece of low, swampy ground overgrown with bushes. Think of the opportunity this combination offers to the Eastern mosquito, and consider my emotions when I found but a solitary individual, and even that one disposed to coquette with me.

I had hidden myself, and was keeping motionless, in order to see the very shy owners of a nest I had found, when the lonely mosquito came as far as the rim of my shade hat, and hovered there, evidently meditating an attack—a mosquito hesitating! I could not stir a hand, or even shake my leafy twig; but it did not require such violent measures; a light puff of breath this side or that was enough to discourage the gentle creature, and in all the hours I sat there it never once came any nearer. The race increased, however, and became rather troublesome on the veranda after tea; but in the grove they were never annoying; I rarely saw half a dozen. When I remember the tortures endured in the dear old woods of the East, in spite of "lollicopop" and pennyroyal, and other horrors with which I have tried to repel them, I could almost decide to live and die in Colorado.

The morning bird chorus in the cottonwood grove where I spent my June was a great shock to me. If my tent had been pitched near the broad plains in which the meadow-lark delights, I might have wakened to the glorious song of this bird of the West. It is not a chorus, indeed, for one rarely hears more than a single performer, but it is a solo that fully makes up for want of numbers, and amply satisfies the lover of bird music, so strong, so sweet, so moving are his notes.

But on my first morning in the grove, what was my dismay—I may almost say despair—to find that the Western wood-pewee led the matins! Now, this bird has a peculiar voice. It is loud, pervasive, and in quality of tone not unlike our Eastern phœbe, lacking entirely the sweet plaintiveness of our Western wood-pewee. A pewee chorus is a droll and dismal affair. The poor things do their best, no doubt, and they cannot prevent the pessimistic effect it has upon us. It is rhythmic, but not in the least musical, and it has a weird power over the listener. This morning hymn does not say, as does the robin's, that life is cheerful, that another glorious day is dawning. It says, "Rest is over; another day of toil is here; come to work." It is monotonous as a frog chorus, but there is a merry thrill in the notes of the amphibian which are entirely wanting in the song. If it were not for the light-hearted tremolo of the chewink thrown in now and then, and the loud, cheery ditty of the summer yellow-bird, who begins soon after the pewee, one would be almost superstitious about so unnatural a greeting to the new day. The evening call of the bird is different. He will sit far up on a dead twig of an old pine-tree, and utter a series of four notes, something like "do, mi, mi, do," repeating them without pausing till it is too dark to see him, all the time getting lower, sadder, more deliberate, till one feels like running out and committing suicide or annihilating the bird of ill-omen.

I felt myself a stranger indeed when I reached this pleasant spot, and found that even the birds were unfamiliar. No robin or bluebird greeted me on my arrival; no cheerful song-sparrow tuned his little pipe for my benefit; no phœbe shouted the beloved name from the peak of the barn. Everything was strange. One accustomed to the birds of our Eastern States can hardly conceive of the country without robins in plenty; but in this unnatural corner of Uncle Sam's dominion I found but one pair.

The most common song from morning till night was that of the summer yellow-bird, or yellow warbler. It was not the delicate little strain we are accustomed to hear from this bird, but a loud, clear carol, equal in volume to the notes of our robin. These three birds, with the addition of a vireo or two, were our main dependence for daily music, though we were favored occasionally by others. Now the Arkansas goldfinch uttered his sweet notes from the thick foliage of the cottonwood-trees; then the charming aria of the catbird came softly from the tangle of rose and other bushes; the black-headed grosbeak now and then saluted us from the top of a pine-tree; and rarely, too rarely, alas! a passing meadow-lark filled all the grove with his wonderful song.

And there was the wren! He interested me from the first; for a wren is a bird of individuality always, and his voice reminded me, in a feeble way, of the witching notes of the winter wren, the

"Brown wren from out whose swelling throat
Unstinted joys of music float."

This bird was the house wren, the humblest member of his musical family; but there was in his simple melody the wren quality, suggestive of the thrilling performances of his more gifted relatives; and I found it and him very pleasing.

The chosen place for his vocal display was a pile of brush beside a closed-up little cottage, and I suspected him of having designs upon that two-roomed mansion for nesting purposes. After hopping all about the loose sticks, delivering his bit of an aria a dozen times or more, in a most rapturous way, he would suddenly dive into certain secret passages among the dead branches, when he was instantly lost to sight. Then, in a few seconds, a close watcher might sometimes see him pass like a shadow, under the cottage, which stood up on corner posts, dart out the farther side, and fly at once to the eaves.

One day I was drawn from the house by a low and oft-repeated cry, like "Hear, hear, hear!" It was emphatic and imperative, as if some unfortunate little body had the business of the world on his shoulders, and could not get it done to his mind. I carefully approached the disturbed voice, and was surprised to find it belonged to the wren, who was so disconcerted at sight of me, that I concluded this particular sort of utterance must be for the benefit of his family alone. Later, that kind of talk, his lord-and-master style as I supposed, was the most common sound I heard from him, and not near the cottage and the brush heap, but across the brook. I thought that perhaps I had displeased him by too close surveillance, and he had set up housekeeping out of my reach. Across the brook I could not go, for between "our side" and the other raged a feud, which had culminated in torn-up bridges and barbed wire protections.

One day, however, I had a surprise. In studying another bird, I was led around to the back of the still shut-up cottage, and there I found, very unexpectedly, an exceedingly busy and silent wren. He did sing occasionally while I watched him from afar, but in so low a tone that it could not be heard a few steps away. Of course I understood this unnatural circumspection, and on observing him cautiously, I saw that he made frequent visits to the eaves of the cottage, the very spot I had hoped he would nest. Then I noted that he carried in food, and on coming out he alighted on a dead bush, and sang under his breath. Here, then, was the nest, and all his pretense of scolding across the brook was but a blind! Wary little rogue! Who would ever suspect a house wren of shyness?

I had evidently done him injustice when I regarded the scolding as his family manner, for here in his home he was quiet as a mouse, except when his joy bubbled over in trills.

To make sure of my conclusions I went close to the house, and then for the first time (to know it) I saw his mate. She came with food in her beak, and was greatly disturbed at sight of her uninvited guest. She stood on a shrub near me fluttering her wings, and there her anxious spouse joined her, and fluttered his in the same way, uttering at the same time a low, single note of protest.

On looking in through the window, I found that the cottage was a mere shell, all open under the eaves, so that the birds could go in and out anywhere. The nest was over the top of a window, and the owner thereof ran along the beam beside it, in great dudgeon at my impertinent staring. Had ever a pair of wrens quarters so ample,—a whole cottage to themselves? Henceforth, it was part of my daily rounds to peep in at the window, though I am sorry to say it aroused the indignation of the birds, and always brought them to the beam nearest me, to give me a piece of their mind.

Bird babies grow apace, and baby wrens have not many inches to achieve. One day I came upon a scene of wild excitement: two wrenlings flying madly about in the cottage, now plump against the window, then tumbling breathless to the floor, and two anxious little parents, trying in vain to show their headstrong offspring the way they should go, to the openings under the eaves which led to the great out-of-doors. My face at the window seemed to be the "last straw." A much-distressed bird came boldly up to me behind the glass, saying by his manner—and who knows but in words?—"How can you be so cruel as to disturb us? Don't you see the trouble we are in?" He had no need of Anglo-Saxon (or even of American-English!). I understood him at once; and though exceedingly curious to see how they would do it, I had not the heart to insist. I left them to manage their willful little folk in their own way.

The next morning I was awakened by the jolliest wren music of the season. Over and over the bird poured out his few notes, louder, madder, more rapturously than I had supposed he could. He had guided his family safely out of their imprisoning four walls, I was sure. And so I found it when I went out. Not a wren to be seen about the house, but soft little "churs" coming from here and there among the shrubbery, and every few minutes a loud, happy song proclaimed that wren troubles were over for the summer. Far in among the tangle of bushes and vines, I came upon him, as gay as he had been of yore:—

> "Pausing and peering, with sidling head,
> As saucily questioning all I said;
> While the ox-eye danced on its slender stem,
> And all glad Nature rejoiced with them."

The chewink is a curious exchange for the robin. When I noticed the absence of the red-breast, whom—like the poor—we have always with us (at the East), I was pleased, in spite of my fondness for him, because, as every one must allow, he is sometimes officious in his attentions, and not at all reticent in expressing his opinions. I did miss his voice in the morning chorus,—the one who lived in the grove was not much of a singer,—but I was glad to know the chewink, who was almost a stranger. His peculiar trilling song was heard from morning till night; he came familiarly about the camp, eating from the dog's dish, and foraging for crumbs at the kitchen door. Next to the Western wood-pewee, he was the most friendly of our feathered neighbors.

He might be seen at any time, hopping about on the ground, one moment picking up a morsel of food, and the next throwing up his head and bursting into song:—

> "But not for you his little singing,
> Soul of fire its flame is flinging,
> Sings he for himself alone,"

as was evident from the unconscious manner in which he uttered his notes between two mouthfuls, never mounting a twig or making a "performance" of his music. I have watched one an hour at a time, going about in his jerky fashion, tearing up the ground and searching therein, exactly after the manner of a scratching hen. This, by the way, was a droll operation, done with both feet together, a jump forward and a jerk back of the whole body, so rapidly one could hardly follow the motion, but throwing up a shower of dirt every time. He had neither the grace nor the dignity of our domestic biddy.

Matter of fact as this fussy little personage was on the ground, taking in his breakfast and giving out his song, he was a different bird when he got above it. Alighting on the wren's brush heap, for instance, he would bristle up, raising the feathers on head and neck, his red eyes glowing eagerly, his tail a little spread and standing up at a sharp angle, prepared for instant fight or flight, whichever seemed desirable.

I was amused to hear the husky cry with which this bird expresses most of his emotions,—about as nearly a "mew," to my ears, as the catbird executes. Whether frolicking with a comrade among the bushes, reproving a too inquisitive bird student, or warning the neighborhood against some monster like a stray kitten, this one cry seemed to answer for all his needs, and, excepting the song, was the only sound I heard him utter.

Familiar as the chewink might be about our quarters, his own home was well hidden, on the rising ground leading up to the mesa,—

"An unkempt zone,
Where vines and weeds and scrub oaks intertwine,"

which no one bigger than a bird could penetrate. Whenever I appeared in that neighborhood, I was watched and followed by anxious and disturbed chewinks; but I never found a nest, though, judging from the conduct of the residents, I was frequently "very warm" (as the children say).

About the time the purple aster began to unclose its fringed lids, and the mariposa lily to unfold its delicate cups on the lower mesa,—nearly the middle of July, —full-grown chewink babies, in brown coats and streaked vests, made their appearance in the grove, and after that the whole world might search the scrub oaks and not a bird would say him nay.

"All is silent now
Save bell-note from some wandering cow,
Or rippling lark-song far away."

III. AN UPROAR OF SONG

The bird music of Colorado, though not so abundant as one could wish, is singularly rich in quality, and remarkable for its volume. At the threshold of the State the traveler is struck by this peculiarity. As the train thunders by, the Western meadow-lark mounts a telegraph pole and pours out such a peal of melody that it is distinctly heard above the uproar of the iron wheels.

This bird is preëminently the bird of the mesa, or high table-land of the region, and only to hear his rare song is well worth a journey to that distant wonderland. Not of his music could Lucy Larcom say, as she so happily does of our bird of the meadow,—

"Sounds the meadow-lark's refrain
Just as sad and clear."

Nor could his sonorous song be characterized by Clinton Scollard's exquisite verse,—

"From whispering winds
your plaintive notes were drawn."

For the brilliant solo of Colorado's bird is not in the least
like the charming minor chant of our Eastern lark. So
powerful that it is heard at great distances in the clear
air, it is still not in the slightest degree strained or harsh,
but is sweet and rich, whether it be close at one's side in
the silence, or shouted from the housetop in the tumult
of a busy street. It has, moreover, the same tender win-
someness that charms us in our own lark song; something that fills the sympathetic listener with delight, that
satisfies his whole being; a siren strain that he longs to
listen to forever. The whole breadth and grandeur of the
great West is in this song, its freedom, its wildness, the
height of its mountains, the sweep of its rivers, the
beauty of its flowers,—all in the wonderful performance.
Even after months of absence, the bare memory of the
song of the mesa will move its lover to an almost painful
yearning. Of him, indeed, Shelley might truthfully say,—

"Better than all measures
Of delightful sound,
Better than all treasures
That in books are found,
Thy skill to poet were,
Thou scorner of the ground."

Nor is the variety of the lark song less noteworthy
than its quality. That each bird has a large répertoire I
cannot assert, for my opportunities for study have been
too limited; but it is affirmed by those who know him
better, that he has, and I fully believe it.

One thing is certainly true of nearly if not quite all of our native birds, that no two sing exactly alike, and the close observer soon learns to distinguish between the robins and the song-sparrows of a neighborhood, by their notes alone. The Western lark seems even more than others to individualize his utterances, so that constant surprises reward the discriminating listener. During two months of bird-study in that delightful cañon-hidden grove at the foot of Cheyenne Mountain, one particular bird song was for weeks an unsolved mystery. The strain consisted of three notes in loud, ringing tones, which syllabled themselves very plainly in my ear as "Whip-for-her."

This unseemly, and most emphatic, demand came always from a distance, and apparently from the top of some tall tree, and it proved to be most tantalizing; for although the first note invariably brought me out, opera-glass in hand, I was never able to come any nearer to a sight of the unknown than the sway of a twig he had just left.

One morning, however, before I was up, the puzzling songster visited the little grove under my windows, and I heard his whole song, of which it now appeared the three notes were merely the conclusion. The performance was eccentric. It began with a soft warble, apparently for his sole entertainment, then suddenly, as if overwhelmed by memory of wrongs received or of punishment deserved, he interrupted his tender melody with a loud, incisive "Whip-for-her!" in a totally different manner. His nearness, however, solved the mystery; the ring of the meadow-lark was in his tones, and I knew him at once. I had not suspected his identity, for the

Western bird does not take much trouble to keep out of sight, and, moreover, his song is rarely less than six or eight notes in length.

Another unique singer of the highlands is the horned lark. One morning in June a lively carriage party passing along the mountain side, on a road so bare and bleak that it seemed nothing could live there, was startled by a small gray bird, who suddenly dashed out of the sand beside the wheels, ran across the path, and flew to a fence on the other side. Undisturbed, perhaps even stimulated, by the clatter of two horses and a rattling mountain wagon, undaunted by the laughing and talking load, the little creature at once burst into song, so loud as to be heard above the noisy procession, and so sweet that it silenced every tongue.

"How exquisite! What is it?" we asked each other, at the end of the little aria.

"It's the gray sand bird," answered the native driver.

"Otherwise the horned lark," added the young naturalist, from his broncho behind the carriage.

Let not his name mislead: this pretty fellow, in soft, gray-tinted plumage, is not deformed by "horns;" it is only two little tufts of feathers, which give a certain piquant, wide-awake expression to his head, that have fastened upon him a title so incongruous. The nest of the desert-lover is a slight depression in the barren earth, nothing more; and the eggs harmonize with their surroundings in color. The whole is concealed by its very openness, and as hard to find, as the bobolink's cradle in the trackless grass of the meadow.

Most persistent of all the singers of the grove beside the house was the yellow warbler, a dainty bit of featherhood the size of one's thumb. On the Atlantic coast his simple ditty is tender, and so low that it must be listened for; but in that land of "skies so blue they flash," he sings it at the top of his voice, louder than the robin song as we know it, and easily heard above the roar of the wind and the brawling of the brook he haunts.

Before me at this moment is the nest of one of these little sprites, which I watched till the last dumpy infant had taken flight, and then secured with the branchlet it was built upon. It was in a young oak, not more than twelve feet from the ground, occupying a perpendicular fork, where it was concealed and shaded by no less than sixteen twigs, standing upright, and loaded with leaves. The graceful cup itself, to judge by its looks, might be made of white floss silk,—I have no curiosity to know the actual material,—and is cushioned inside with downy fibres from the cottonwood-tree. It is dainty enough for a fairy's cradle.

The Western wood-pewee, in dress and manners nearly resembling his Eastern brother.

"The pewee of the loneliest woods,
  Sole singer in the solitudes,"

has a strange and decidedly original utterance. While much louder and more continuous, it lacks the sweetness of our bird's notes; indeed, it resembles in quality of tone the voice of our phœbe, or his beautiful relative, the great-crested flycatcher. The Westerner has a great deal to say for himself. On alighting, he announces the fact by a single note, which is a habit also of our phœbe; he sings the sun up in the morning, and he sings it down in the evening, and he would be a delightful neighbor if only his voice were pleasing. But there is little charm in the music, for it is in truth a dismal chant, with the air and cheerfulness of a funeral dirge—a pessimistic performance that inspires the listener with a desire to choke him then and there.

This bird's nest, as well as his song, is unlike that of our Western wood-pewee. Instead of a delicate, lichen-covered saucer set lightly upon a horizontal crotch of a dead branch,—our bird's chosen home,—it is a deeper cup, fastened tightly upon a large living branch, and, at least in a cottonwood grove, decorated on the outside with the fluffy cotton from the trees.

Even the humming-bird, who contents himself in this part of the world with a modest hum, heard but a short distance away, at the foot of the Rocky Mountains may almost be called a noisy bird. The first one I noticed dashed out of a thickly leaved tree with loud, angry cries, swooped down toward me, and flew back and forth over my head, scolding with a hum which, considering his size, might almost be called a roar. I could not believe my ears until my eyes confirmed their testimony. The sound was not made by the wings, but was plainly a cry strong and harsh in an extraordinary degree.

The Western ruby-throat has other singularities which differentiate him from his Eastern brother. It is very droll to see one of his family take part in the clamors of a bird mob, perching like his bigger fellows, and adding his excited cries to the notes of catbird and robin, chewink and yellow-bird. Attracted one morning by a great bird outcry in a dense young oak grove across the road, I left my seat under the cottonwoods and strolled over toward it. It was plain that some tragedy was in the air, for the winged world was in a panic. Two robins, the only pair in the neighborhood, uttered their cry of distress from the top of the tallest tree; a catbird hopped from branch to branch, flirting his tail and mewing in agitation; a chewink or two near the ground jerked themselves about uneasily, adding their strange, husky call to the hubbub; and above the din rose the shrill voice of a humming-bird. Every individual had his eyes fixed upon the ground, where it was evident that some monster must be lurking. I expected a big snake at the very least, and, putting the lower branches aside, I, too, peered into the semi-twilight of the grove.

No snake was there; but my eyes fell upon an anxious little gray face, obviously much disturbed to find itself the centre of so much attention. As I appeared, this bugaboo, who had caused all the excitement, recognized me as a friend and ran toward me, crying piteously. It was a very small lost kitten!

I took up the stray little beastie, and a silence fell upon the assembly in the trees, which began to scatter, each one departing upon his own business in a moment. But the humming-bird refused to be so easily pacified; he was bound to see the end of the affair, and he followed me out of the grove, still vigorously speaking his mind about the enemy in fur. I suspected that the little creature had wandered away from the house on the hill above, and I went up to see. The hummer accompanied me every step of the way, sometimes flying over my head, and again alighting for a minute on a branch under which I passed. Not until he saw me deliver pussy into the hands of her own family, and return to my usual seat in the grove, did he release me from surveillance and take his leave.

The yellow-breasted chat, the long-tailed variety belonging to the West, delivers his strange medley of "chacks" and whistles, and rattles and other indescribable cries, in a voice that is loud and distinct, as well as sweet and rich. He is a bird of humor, too, with a mocking spirit not common in his race. One day, while sitting motionless in a hidden nook, trying to spy upon the domestic affairs of this elusive individual, I was startled by the so-called "laugh" of a robin, which was instantly repeated by a chat, unseen, but quite near. The robin, apparently surprised or interested, called again, and was a second time mocked. Then he lost his temper, and began a serious reproof to the levity of his neighbor, which ended in a good round scolding, as the saucy chat continued to repeat his taunting laugh. This went on till the red-breast flew away in high dudgeon.

Why our little brothers in feathers are so much more boisterous than elsewhere,

> "Up in the parks and the mesas wide,
>   Under the blue of the bluest sky,"

has not, so far as I know, been discovered.

Whether it be the result of habitual opposition to the strong winds which, during the season of song, sweep over the plains every day, or whether the exhilaration of the mountain air be the cause—who can tell?

IV. THE TRAGEDY OF A NEST

Near to the Camp, a little closer to beautiful Cheyenne Mountain, lay a small park. It was a continuation of the grove, through which the brook came roaring and tumbling down from the cañons above, and, being several miles from the town, it had never become a popular resort. A few winding paths, and a rude bench here and there, were the only signs of man's interference with its native wildness; it was practically abandoned to the birds—and me.

The birds had full possession when I appeared on the scene, and though I did my best to be unobtrusive, my presence was not so welcome as I could have wished. Every morning when I came slowly and quietly up the little path from the gate, bird-notes suddenly ceased; the grosbeak, pouring out his soul from the top of a pine-tree, dived down the other side; the towhee, picking up his breakfast on the ground, scuttled behind the bushes and disappeared; the humming-bird, interrupted in her morning "affairs," flew off over my head, scolding vigorously; only the vireo—serene as always—went on warbling and eating, undisturbed.

Then I made haste to seek out an obscure spot, where I could sit and wait in silence, to see who might unwittingly show himself.

I was never lonely, and never tired; for if—as sometimes happened—no flit of wing came near to interest me, there before me was beautiful Cheyenne, with its changing face never twice alike, and its undying associations with its poet and lover, whose lonely grave makes it forever sacred to those who loved her. There, too, was the wonderful sky of Colorado, so blue it looked almost violet, and near at hand the "Singing Water," whose stirring music was always inspiring.

One morning I was startled from my reverie by a sudden cry, so loud and clear that I turned quickly to see what manner of bird had uttered it. The voice was peculiar and entirely new to me. First came a scolding note like that of an oriole, then the "chack" of a blackbird, and next a sweet, clear whistle, one following the other rapidly and vehemently, as if the performer intended to display all his accomplishments in a breath. Cheyenne vanished like "the magic mountain of a dream," blue skies were forgotten, the babbling brook unheard, every sense was instantly alert to see that extraordinary bird,—

> "Like a poet hidden,
> Singing songs unbidden."

But he did not appear. Not a leaf rustled, not a twig bent, though the strange medley kept on for fifteen minutes, then ceased as abruptly as it had begun, and not a whisper more could be heard. The whole thing seemed uncanny. Was it a bird at all, or a mere "wandering voice"? It seemed to come from a piece of rather swampy ground, overgrown with clumps of willow and low shrubs; but what bird of earthly mould could come and go, and make no sign that a close student of bird ways could detect? Did he creep on the ground? Did he vanish into thin air?

Hours went by. I could not go, and my leafy nook was "struck through with slanted shafts of afternoon" before I reluctantly gave up that I should not see my enchanter that day, and slowly left the grove, the mystery unexplained.

Very early the next morning I was saluted by the same loud, clear calls near the house. Had then the Invisible followed me home? I sprang up and hurried to the always open window. The voice was very near; but I could not see its author, though I was hidden behind blinds.

This time the bird—if bird it were—indulged in a fuller répertoire. I seized pencil and paper, and noted down phonetically the different notes as they were uttered. This is the record: "Rat-t-t-t-t" (very rapid); "quit! quit! quit!" (a little slower); "wh-eu! wh-eu!" (still more deliberately); "chack! chack! chack!" (quite slow); "cr[ee], cr[ee], cr[ee]" (fast); "hu-way! hu-way!" (very sweet). There was a still more musical clause that I cannot put into syllables, then a rattle exactly like castanets, and lastly a sort of "kr-r-r! kr-r-r!" in the tone of a great-crested flycatcher. While this will not express to one who has not heard it the marvelous charm of it all, it will at least indicate the variety.

Hardly waiting to dispose of breakfast, I betook myself to my "woodland enchanted," resolved to stay till I saw that bird.

> "All day in the bushes
> The woodland was haunted."

The voice was soon on hand, and once more I was treated to the incomparable recitative.

This day, too, my patience was rewarded; the mystery was solved; I saw the Unknown! While my eyes were fixed upon a certain bush before me, the singer incautiously ventured too near the top of a twig, and I saw him plainly, standing almost upright, and vehemently chanting his fantasia, opening his mouth very wide with every call. I knew him at once, the rogue! from having read of him; he was the yellow-breasted chat. It was well, indeed, that I happened to be looking at that very spot, and that I was quick in my observation; for in a moment he saw the blunder he had made, and slipped back down the stem, too late for his secret—I had him down in black and white.

From that time the little park was never lonely, nor did I spend much time dreaming over Cheyenne. The moment I appeared in the morning my lively host began his vocal gymnastics, while I sat spellbound, bewitched by the magic of his notes. In spite of being absorbed in listening to him, I retained my faculties sufficiently to reflect that the chat had probably other employment than entertaining me, and that doubtless his object was to distract my attention from looking about me, or to reproach me for intruding upon his private domain. In either case there was, of course,

> "A nest unseen
> Somewhere among the million stalks;"

and, delightful as I found the unseen bird, his nest was a treasure I was even more anxious to see.

Not to disturb him more than necessary, I spent part of an evening studying up the nesting habits of the chat,—the long-tailed, yellow-breasted, as I found him to be,—and the next morning made a thorough search through the swamp, looking into every bush and examining every thicket. An hour or two of this hard work satisfied me for the day, and I went home warm and tired, followed to the very door by the mocking voice, triumphing, as it seemed, in my failure.

The next day, however, fortune smiled upon me; I came upon a nest, not far above the ground, among the stems of a clump of shrubs, which exactly answered the description of the one I sought. Careful not to lay a finger on it, I slightly parted the branches above, and looked in upon three pinkish-white eggs, small in size and dainty as tinted pearls. Happy day, I thought, and the forerunner of happy to-morrows when I should watch

> "The green nest full of pleasant shade
> Wherein three speckled eggs were laid,"

and see and delight in the family life centring about it.

To study a bird so shy required extraordinary precautions; I therefore sought, and found, a post of observation a long way off, where I could look through a natural vista among the shrubs, and with my glass bring the

bush and its precious contents into view. For greater seclusion in my retreat, so that I should be as little conspicuous as possible, I drew down a branch of the low tree
over my seat, and fastened it with a fine string to a stout
weed below. Then I thought I had a perfect screen; I de-
voutly hoped the birds would not notice me.

Vain delusion! and labor as vain! Doubtless two
pairs of anxious eyes watched from some neighboring
bush all my careful preparations, and then and there two
despairing hearts bade farewell to their lovely little
home, abandoned it and its treasures to the spy and the
destroyer, which in their eyes I seemed to be.

This conclusion was forced upon me by the ex-
periences of the next few days. The birds absolutely
would not approach the nest while I was in the park. The
first morning I sat motionless for nearly two hours, and
not a feather showed itself near that bush; it was plainly
"tabooed." During the next day the chat called from this
side and that, moving about in his wonderful way, without disturbing a twig, rustling a leaf, or flitting a wing—
as silently, indeed, as if he were a spirit unclothed.

While waiting for him to show himself, making
myself as nearly a part of nature about me as a mortal is
gifted to do, I congratulated myself upon the one good
look I had secured, for, with all my efforts and all my
watching, I saw him but twice more all summer. The
enigma of that remarkable voice would have been maddening indeed, if I could not have known to whom it
belonged.

After several days of untiring observation I had
but two glimpses to record. On one occasion a chat alighted on the top sprig of the fateful shrub, as if going to the

nest, but almost on the instant vanished. The same day, a little later, one of these birds flitted into my view, without a sound. So perfectly silent were his movements that I should not have seen him if he had not come directly before my eyes. He, or she, for the pair are alike, alighted in a low bush and scrambled about as if in search of insects, climbing, not hopping. He stayed but a few seconds and departed like a shadow, as he had come.

On the tenth day after my discovery of the nest with its trio of eggs I went out as usual, for I could not abandon hope. In passing the nest I glanced in and saw one egg; I could never see but one as I went by, but, not liking to go too near, I presumed that the other two were there, as I had always found them, and slipped quietly into my usual place.

In a few moments the chat shouted a call so near that it fairly startled me. From that he went on to make his ordinary protest, but, as happened nearly every time, I was not able to see him. I saw something—something that took my breath away. A shadowy form creeping stealthily through the shrubs five or six feet from me. It glided across the opening in front, and in a moment went to the bush I was watching. In silence, but with evident excitement, it moved about, approached the nest, and in a few seconds flew quickly across the path in plain sight, holding in its mouth something white which was large for its beak. I was reminded of an English sparrow carrying a piece of bread as big as his head, a sight familiar to every one. In a minute or two the same bird, or his twin, came to the nest again and disappeared on the other side.

When I left my place to go home, I looked with misgivings into the nest on which I had built so many hopes. Lo! it was empty!

Now I identified that stealthy visitor absolutely, but I shall never name him. I have never heard him accused of nest-robbing, and I shall not make the charge; for I am convinced that the chat had deserted the nest, and that this abstracter of eggs knew it, and simply took the good things the gods threw in his way—as would the best of us.

After that unfortunate ending the chat disappeared from the little park; but a week later I came upon him, or his voice, in a private and rarely visited pasture down the road, where many clumps of small trees and much low growth offered desirable nesting-places. He made his usual protest, and feeling that I had been the cause of the tragedy of the first nest, though I had grieved over it as much as the owners could, the least I could do, to show my regret, was to take myself and my curiosity out of his neighborhood. So I retired at once, and left the whole broad pasture to the incorrigible chat family, who, I hope, succeeded at last in enriching the world by half a dozen more of their bewitching kind.

V. A FEAST OF FLOWERS

When first the crocus thrusts its point of gold
Up through the still snow-drifted garden mould,
And folded green things in dim woods unclose
Their crinkled spears, a sudden tremor goes
Into my veins and makes me kith and kin
To every wild-born thing that thrills and blows.

T. B. Aldrich

My feast of flowers began before I entered
Colorado. For half the breadth of Kansas the banks of the
railroad were heavenly blue with clustered blossoms of
the spiderwort. I remember clumps of this flower in my
grandmother's old-fashioned garden, but my wildest
dreams never pictured miles of it, so profuse that, looking backward from the train, the track looked like
threads of steel in a broad ribbon of blue.

Through the same State, as well, the Western
meadow-larks kept us company, and I shall never again

think of "bleeding Kansas," but of smiling Kansas, the home of the bluest of blossoms and the sweetest of singers. The latter half of the way through the smiling State was golden with yellow daisies in equal abundance, and beside them many other flowers. Beginning at noon, I counted twenty-seven varieties, so near the track that I could distinguish them as we rushed past.

The Santa Fé road enters Colorado in a peculiarly desolate region. Flowers and birds appear to have stayed behind in Kansas, and no green thing shows its head, excepting one dismal-looking bush, which serves only to accentuate the poverty of the soil. As we go on, the mud is replaced by sand and stones, from gravel up to big bowlders, and flowers begin to struggle up through the unpromising ground.

Nothing is more surprising than the amazing profusion of wild-flowers which this apparently ungenial soil produces. Of a certainty, if Colorado is not the paradise of wild-flowers, it is incomparably richer in them than any State east of the Mississippi River and north of "Mason and Dixon's Line." To begin with, there is a marvelous variety. Since I have taken note of them, from about the 10th of June till nearly the same date in July, I have found in my daily walk of not more than a mile or two, each time from one to seven new kinds. A few days I have found seven, many times I have brought home four, and never has a day passed without at least one I had not seen before. That will average, at a low estimate, about a hundred varieties of flowers in a month, and all within a radius of four miles. What neighborhood can produce a record equal to this?

Then, again, the blossoms themselves are so abundant. Hardly a root contents itself with a single flower. The moccasin-plant is the only one I have noticed as yet. One root will usually send up from one to a dozen stems, fairly loaded with buds—like the yucca—which open a few every day, and thus keep in bloom for weeks. Or if there is but one stem, it will be packed with buds from the ground to the tip, with new ones to come out for every blossom that falls.

One in the vase on my stand at this moment is of this sort. It is a stem that sometimes attains a height of four or five feet. I think it lengthens as long as it is blossoming, and, to look at its preparations, that must be all summer. Every two or three inches of the stout stem is a whorl of leaves and buds and blossoms. Except the number of buds, it is all in fours. Opposite each other, making a cross, are four leaves, like a carnation leaf at first, but broadening and lengthening till it is two inches at the base and eight or ten long. Rising out of the axil of each leaf are buds, of graduated size and development up to the open blossom. That one stem, therefore, is prepared to open fresh flowers every day for a long time.

The plant is exquisitely beautiful, for the whole thing, from the stem to the flower petals, is of a delicate, light pea-green. The blossom opens like a star, with four stamens and four petals. The description sounds mathematical, but the plant is graceful—a veritable symphony in green.

A truly royal bouquet stands on my table—three spikes of yucca flowers in a tall vase, the middle one three feet high, bearing fifty blossoms and buds, of large size and a pink color; on its right, one a little less in size, with long creamy cups fully open; and on the left another, set with round greenish balls, not so open as cups. They are distinctly different, but each seems more exquisite than the other, and their fragrance fills the room. In fact it is so overpowering that when at night I close the door opening into the grove, I shut the vase and its contents outside.

This grand flower is the glory of the mesa or table-land at the foot of this range of the Rocky Mountains —the Cheyenne Range. Where no grass—that we name grass—will grow, where trees die for want of water, these noble spikes of flowers dot the bare plains in profusion.

It is the rich possessor of three names. To the flower-lover it is the yucca; to the cultivator, or whosoever meddles with its leaves, it is the Spanish-bayonet; to the utilitarian, who values a thing only as it is of use to him, it is the soap-weed—ignoble name, referring to certain qualities pertaining to its roots. When we remember that this flower is not the careful product of the garden, but a product of spontaneous growth in the most barren and hopeless-looking plains, we may well regard it as a type of Colorado's luxuriance in these loveliest of nature's gifts.

Of a surly disposition is the blossom of a cactus— the "prickly-pear," as we call it in Eastern gardens, where we cultivate it for its oddity, I suppose. When the sojourner in this land of flowers sees, opening on all sides of this inhospitable-looking plant, rich cream-colored cups, the size of a Jacqueminot bud, and of a rare, satiny sheen, she cannot resist the desire to fill a low dish with them for her table.

Woe to her if she attempts to gather them "by hand"! Properly warned, she will take a knife, sever the flower from the pear (there is no stem to speak of), pick it up by the tip of a petal, carry it home in a paper or handkerchief, and dump it gently into water—happy if she does not feel a dozen intolerable prickles here and there, and have to extract, with help of magnifying-glass and tweezers, as many needle-like barbs rankling in her flesh. She may as well have spared herself the trouble. The flowers possess the uncompromising nature of the stock from which they sprung; they will speedily shut themselves up like buds again—I almost believe they close with a snap—and obstinately refuse to display their satin draperies to delight the eyes of their abductors. This unlovely spirit is not common among Colorado flowers; most of them go on blooming in the vase day after day.

Remarkable are the places in which the flowers are found. Not only are they seen in crevices all the way up the straight side of rocks, where one would hardly think a seed could lodge, but beside the roads, between the horses' tracks, and on the edge of gutters in the streets of a city. One can walk down any street in Colorado Springs and gather a bouquet, lovely and fragrant, choice enough to adorn any one's table. I once counted twelve varieties in crossing one vacant corner lot on the principal street.

One of the richest wild gardens I know is a bare, open spot in a cottonwood grove, part of it tunneled by ants, which run over it by millions, and the rest a jumble of bowlders and wild rosebushes, impossible to describe. In this spot, unshaded from the burning sun, flourish flowers innumerable. Rosebushes, towering far above one's head, loaded with bloom; shrubs of several kinds, equally burdened by delicate white or pink blossoms; the ground covered with foot-high pentstemons, blue and lavender, in which the buds fairly get in each other's way; and a curious plant—primrose, I believe—which opens every morning, a few inches from the ground, a large white blossom like the magnolia, turns it deep pink, and closes it before night; several kinds of yellow flowers; wild geraniums, with a look of home in their daintily penciled petals; above all, the wonderful golden columbine. I despair of picturing this grand flower to eyes accustomed to the insignificant columbine of the East. The blossom is three times the size of its Eastern namesake, growing in clumps sometimes three feet across, with thirty or forty stems of flowers standing two and a half feet high. In hue it is a delicate straw color, sometimes all one tint, sometimes with outside petals of snowy white, and rarely with those outsiders of lavender. It is a red-letter day when the flower-lover comes upon a clump of the lavender-leaved columbine. Far up in the mountains is found still another variety of this beautiful flower, with outside petals of a rich blue. This, I believe, is the State flower of Colorado.

I am surprised at the small number of flowers here with which I am familiar. I think there are not more than half a dozen in all this extraordinary "procession of flowers" that I ever saw before. In consequence, every day promises discoveries, every walk is exciting as an excursion into unknown lands, each new find is a fresh treasure.

VI. A CINDERELLA AMONG FLOWERS

Like torches lit for carnival,
The fiery lilies straight and tall
Burn where the deepest shadow is;
Still dance the columbines cliff-hung,
And like a broidered veil outflung
The many-blossomed clematis.

Susan Coolidge

A rough, scraggy plant, with unattractive, dark-green foliage and a profusion of buds standing out at all angles, is, in July, almost the only growing thing to be seen on the barren-looking mesa around Colorado Springs. Anything more unpromising can hardly be imagined; the coarsest thistle is a beauty beside it; the common burdock has a grace of growth far beyond it; the meanest weed shows a color which puts it to shame. Yet if the curious traveler pass that way again, late in the afternoon, he shall find that "Solomon in all his glory was not arrayed like one of these." He will see the bush transfigured; its angular form hidden under a mass of many pointed stars of snowy whiteness, with clusters of pale gold stamens. Then will stand revealed the "superb mentzelia," a true Cinderella, fit only for ignominious uses in the morning, but a suitable bride for the fairy prince in the evening.

To look at the wide-stretching table-lands, where, during its season, this fairy-story transformation takes place daily, so burned by the sun, and swept by the wind, that no cultivated plant will flourish on it, one would never suspect that it is the scene of a brilliant "procession of flowers" from spring to fall. "There is always something going on outdoors worth seeing," says Charles Dudley Warner, and of no part of the world is this more true than of these apparently desolate plains at the foot of the Rocky Mountains. Rich is the reward of the daily stroller, not only in the inspiration of its pure, bracing air, the songs of its meadow-larks, and the glory of its grand mountain view, but in its charming flower show.

This begins with the anemone, modest and shy like our own, but three times as big, and well protected from the sharp May breezes by a soft, fluffy silk wrap. Then some day in early June the walker shall note groups of long, sword-shaped leaves, rising in clusters here and there from the ground. He may not handle them with impunity, for they are strong and sharp-edged, and somewhat later the beauty they are set to guard is revealed. A stem or two, heavy and loaded with hard green balls, pushes itself up among them day by day, till some morning he stands spellbound before the full-blown bells of the yucca, cream-tinted or pink, and fragrant as the breath of summer.

Before the Nature-lover is tired of feasting his eyes upon that stately flower, shall begin to unfold the crumpled draperies of the great Mexican poppy, dotting the hillsides and the mesa with white, as far as the eye can reach. Meanwhile, the earth itself shall suddenly turn to pink, and a close look disclose a tiny, low-growing blossom, sweet as the morning, with the glow of the sunrise in its face; a little bunch of crazy-looking stamens, and tiny snips of petals standing out at all angles, and of all shades on one stem, from white to deep red; the whole no bigger than a gauzy-winged fly, and shaped not unlike one, with a delicious odor that scents the air.

Next day—or next week—wandering over the pathless barrens, the observer may come upon a group of cream-colored satin flowers, wide open to the sun, innocent looking and most tempting to gather. But the great fleshy leaves from which they spring give warning; they belong to the cactus family, and are well armed to protect their treasures from the vagrant hand. The walker— if he be wise—will content himself with looking, nor seek a nearer acquaintance.

While these royal beauties are adorning the highlands, others, perhaps even more lovely, are blooming in the cañons, under the trees, and beside the noisy brooks. First, there is a "riot of roses"—the only expression that adequately suggests the profusion of these beautiful flowers. They grow in enormous bushes, far above one's head, in impenetrable thickets, extending for yards each way.

> "Rose hedges
> Abloom to the edges."

Every country road is walled in by them; every brookside is glorified by their rich masses of color; and no rocky wall is so bare but here and there a tiny shoot finds root, and open its rosy bloom. All these bushes, from the low-growing sort that holds its mottled and shaded petals three inches above the ground, to that whose top one cannot reach, are simply loaded with blossoms of all shades, from nearly white to deepest rose-color, filling the air with perfume.

The first time one comes upon this lavish display, he—or more probably she—picks a spray from the first bush; she cannot resist the next variety, and before she knows it her arms are full, with temptations as strong as ever before her. She may at last, like "H. H.," take home her roses by the carriage load, or, overwhelmed by their numbers, leave them all on their stems, and enjoy them in mass.

Shyly hiding under the taller shrubs beside the running water, the experienced seeker will find the gilia, one of the gems of Colorado's bouquet. This plant consists of one slender stem two feet or more tall, swayed by every breeze, and set for several inches of its length with daintiest blossoms,—

> "Like threaded rubies on its stem."

They are like fairy trumpets, in many shades, from snow white to deep rose, and brilliant scarlet, with great variety of delicate marking visible only under a glass. The stem is so sticky that the flowers must be arranged as they are gathered; for they cling to each other more closely than the fabled "brother," and an attempt to separate them will result in torn flowers.

Anything more exquisite than a vase of gilias alone is rarely seen. The buds are as lovely as the blossoms; new ones open every day, and even the faded ones are not unsightly; their petals are simply turned backward a little. One minute every morning spent in snipping off blossoms that are past their prime insures the happy possessor a bouquet that is a joy forever, even in memory; lovely and fresh, in ever-changing combinations of color and form.

Some day shall be made memorable to the enthusiast by the discovery of a flower which should be named for "H. H.,"—the one which looked so charming from the moving train that her winning tongue brought the iron horse to a pause while it was gathered, "root and branch," for her delectation. Finding the gorgeous spike of golden blossoms without a common name, she called it—most happily—the golden prince's feather. It is to be presumed that it has an unwieldy scientific cognomen in the botanies; but I heard of no common one, except that given by the poet.

While this royal flower is still in bloom, may be found the mariposa, or butterfly lily, small and low on the burning mesa, but more generous in size, and richer of hue, in the shaded cañons.

> "Like a bubble borne in air
> Floats the shy Mariposa's bell,"

says Susan Coolidge in her beautiful tribute to her beloved friend and poet. The three petals of this exquisite flower form a graceful cup of differing degrees of violet hue, some being nearly white, with the color massed in a rich, deep-toned crescent, low down at the heart of each petal, while others are glowing in the most regal purple.

All these weeks, too, have been blossoming dozens, yes, hundreds of others; every nook and corner is full; every walk brings surprises. Some of our most familiar friends are wanting. One is not surprised that the most common wayside flower of that golden region is the yellow daisy, or sunflower it is called; but she remembers fondly our fields of white daisies, and clumps of gay little buttercups, and she longs for cheery-faced dandelions beside her path. A few of the latter she may find, much larger and more showy than ours; but these—it is said in Colorado Springs—are all from seed imported by an exile for health's sake, who pined for the flowers of home.

Several peculiarities of Colorado flowers are noteworthy. Some have gummy or sticky stems, like the gilia, already mentioned, and others again are "clinging," by means of a certain roughness of stem and leaf. The mentzelia is of this nature; half a dozen stalks can with difficulty be separated; and they seem even to attract any light substance, like fringe or lace, holding so closely to it that they must be torn apart.

Many of the prettiest flowers are, like our milkweed, nourished by a milky juice, and when severed from the parent stem, not only weep thick white tears, which stain the hands and the garments, but utterly refuse to subsist on water, and begin at once to droop. Is it the vitality in the air which forces even the plants to eccentricities? Or can it be that they have not yet been subdued into uniformity like ours? Are they unconventional—

nearer to wild Nature? So queries an unscientific lover of them all.

This slight sketch of a few flowers gives hardly a hint of the richness of Colorado's flora. No words can paint the profusion and the beauty. I have not here even mentioned some of the most notable: the great golden columbine, the State flower, to which our modest blossom is an insignificant weed;

> "The fairy lilies, straight and tall,
> Like torches lit for carnival;"

the primrose, opening at evening a disk three or four inches across, loaded with richest perfume, and changed to odorless pink before morning; exquisite vetches, with bloom like our sweet pea, and of more than fifty varieties; harebells in great clumps, and castilleias which dot the State with scarlet; rosy cyclamens "on long, lithe stems that soar;" and mertensias, whose delicate bells, blue as a baby's eyes, turn day by day to pink; the cleome, which covers Denver with a purple veil; the whole family of pentstemons, and hundreds of others.

An artist in Colorado Springs, who has given her heart, almost her life, to fixing in imperishable color the floral wealth about her, has painted over three hundred varieties of Colorado wild-flowers, and her list is still incomplete.

It is not pleasant to mar this record of beauty, but one thing must be mentioned. The luxuriance of the flowers is already greatly diminished by the unscrupulousness of the tourists who swarm in the flower season, especially, I am sorry to say, women. Not content with filling their hands with flowers, they fill their arms and even their carriage, if they have one. Moreover, the hold of the plant on the light, sandy soil is very slight; and the careless gatherer, not provided with knife or scissors, will almost invariably pull the root with the flower, thus totally annihilating that plant. When one witnesses such greediness, and remembers that these vandals are in general on the wing, and cannot stay to enjoy what they have rifled, but will leave it all to be thrown out by hotel servants the next morning, he cannot wonder at the indignation of the residents toward the traveler, nor that "No admittance" notices are put up, and big dogs kept, and that "tourist" is a name synonymous with "plunderer," and bitterly hated by the people.

I have seen a party of ladies—to judge by their looks—with arms so full of the golden columbine that it seemed they could not hold another flower, whose traveling dress and equipments showed them to be mere transient passers through, who could not possibly make use of so many. Half a dozen blossoms would have given as much pleasure as half a hundred, and be much more easily cared for, besides leaving a few for their successors to enjoy. The result is, of course, plain to see: a few more years of plunder, and Colorado will be left bare, and lose half her charm.

One beautiful place near Colorado Springs, Glen Eyrie, belonging to General Palmer, was generously left open for every one to enjoy by driving through; but, incredible as it seems, his hospitality was so abused, his lovely grounds rifled, not only of wild-flowers, but even of cultivated flowers and plants, that he was forced at last to put up notices that the public was allowed to "drive through without dismounting."

VII. CLIFF-DWELLERS
IN THE CAÑON

Glad
With light as with a garment it is clad
Each dawn, before the tardy plains have won
One ray; and often after day has long been done
For us, the light doth cling reluctant,
sad to leave its brow.

H. H.

The happiest day of my summer in the Rocky
Mountains was passed in the heart of a mountain conse-
crated by the songs and the grave of its lover, "H. H.,"—
beautiful Cheyenne, the grandest and the most graceful
of its range.

Camp Harding, my home for the season, in its
charming situation, has already been described. The for-
tunate dwellers in this "happy valley" were blessed with

two delectable walks, "down the road" and "up the road." Down the road presented an enchanting procession of flowers, which changed from day to day as the season advanced; to-day the scarlet castilleia, or painter's-brush, flaming out of the coarse grasses; to-morrow the sand lily, lifting its dainty face above the bare sand; next week the harebell, in great clumps, nodding across the field, and next month the mariposa or butterfly lily, just peeping from behind the brush,—with dozens of others to keep them company. As one went on, the fields grew broader, the walls of the mesa lowered and drew apart, till the cañon was lost in the wide, open country.

This was the favorite evening walk, with all the camp dogs in attendance,—the nimble greyhound, the age-stiffened and sedate spaniel, the saucy, ill-bred bull-terrier, and the naïve baby pug. The loitering walk usually ended at the red farmhouse a mile away, and the walkers returned to the camp in the gloaming, loaded with flowers, saturated with the delicious mountain air, and filled with a peace that passeth words.

Up the road led into the mountain, under thick-crowding trees, between frowning rocks, ever growing higher and drawing nearer together, till the carriage road became a burro track, and then a footpath; now this side the boisterous brook, then crossing by a log or two to the other side, and ending in the heart of Cheyenne in a cul-de-sac, whose high perpendicular sides could be scaled only by flights of steps built against the rocks. From high up the mountain, into this immense rocky basin, came the brook Shining Water, in seven tremendous leaps, each more lovely than the last, and reached at bottom a deep stone bowl, which flung it out in a shower of spray forbidding near approach, and keeping the rocks forever wet.

The morning walk was up the road, in the grateful shade of the trees, between the cool rocks, beside the impetuous brook. This last was an ever fresh source of interest and pleasure, for nothing differs more widely from an Eastern brook than its Western namesake. The terms we apply to our mountain rivulets do not at all describe a body of water on its way down a Rocky Mountain valley. It does not murmur,—it roars and brawls; it cannot ripple,—it rages and foams about the bowlders that lie in its path. The name of a Colorado mountain stream, the Roaring Fork, exactly characterizes it.

One warm morning in June, a small party from the camp set out for a walk up the road. By easy stages, resting here and there on convenient rocks, beguiled at every step by something more beautiful just ahead, they penetrated to the end of the cañon. Of that party I was one, and it was my first visit. I was alternately in raptures over the richness of color, the glowing red sandstone against the violet-blue sky, and thrilled by the grandeur of places which looked as if the whole mountain had been violently rent asunder.

But no emotion whatever, no beauty, no sublimity even, can make me insensible to a bird note. Just at the entrance to the Pillars of Hercules, two towering walls of perpendicular rock that approach each other almost threateningly, as if they would close up and crush between them the rash mortal who dared to penetrate farther,—in that impressive spot, while I lingered, half yielding to a mysterious hesitation about entering the strange portal, a bird song fell upon my ear. It was a plaintive warble, that sounded far away up the stern cliff above my head. It seemed impossible that a bird could find a foothold, or be in any way attracted by those bare walls, yet I turned my eyes, and later my glass that way.

At first nothing was to be seen save, part way up the height, an exquisite bit of nature. In a niche that might have been scooped out by a mighty hand, where scarcely a ray of sunlight could penetrate, and no human touch could make or mar, were growing, and blooming luxuriantly, a golden columbine, Colorado's pride and glory, a rosy star-shaped blossom unknown to me, and a cluster of

"Proud cyclamens on long, lithe stems that soar."

When I could withdraw my eyes from this dainty wind-sown garden, I sought the singer, who proved to be a small brown bird with a conspicuous white throat, flitting about on the face of the rock, apparently quite at home, and constantly repeating his few notes. His song was tender and bewitching in its effect, though it was really simple in construction, being merely nine notes, the first uttered twice, and the remaining eight in descending chromatic scale.

Now and then the tiny songster disappeared in what looked like a slight crack in the wall, but instantly reappeared, and resumed his siren strains. Spellbound I stood, looking and listening; but alas! the hour was late, the way was long, and others were waiting; I needs must tear myself away. "To-morrow I will come again," I said, as I turned back. "To-morrow I shall be here alone, and spend the whole day with the cañon wren."

Then we retraced our steps of the morning, lingering among the pleasant groves of cottonwood, oak, and aspen; pausing to admire the cactus display of gorgeous yellow, with petals widespread, yet so wedded to their wildness that they resented the touch of a human hand, resisting their ravisher with needle-like barbs, and then sullenly drawing together their satin petals and refusing to open them more; past great thickets of wild roses, higher than our heads and fragrant as the morning; beside close-growing bushes, where hid the

"Golden cradle of the moccasin flower,"

and the too clever yellow-breasted chat had mocked and defied me; and so home to the camp.

At an early hour the next morning, the carriage of my hostess set me down at the entrance of Cheyenne Cañon proper, with the impedimenta necessary for a day's isolation from civilization. I passed through the gate,—for even this grand work of nature is claimed as private property; but, happily, through good sense or indifference, "improvements" have not been attempted, and one forgets the gate and the gate-keeper as soon as they are passed.

Entering at that unnatural hour, and alone, leaving the last human being behind,—staring in astonishment, by the way, at my unprecedented proceeding,—I began to realize, as I walked up the narrow path, that the whole grand cañon, winding perhaps a mile into the heart of this most beautiful of the Rocky Mountains, was mine alone for three hours. Indeed, when the time arrived for tourists to appear, so little did I concern myself with them that they might have been a procession of spectres passing by; so, in effect, the cañon was my solitary possession for nine blissful hours.

The delights of that perfect day cannot be put into words. Strolling up the path, filled with an inexpressible sense of ownership and seclusion from all the world, I first paused in the neighborhood of the small cliff-dweller whose music had charmed me, and suggested the enchanting idea of spending a day with him in his retreat. I seated myself opposite the forbidding wall where the bird had hovered, apparently so much at home. All was silent; no singer to be heard, no wren to be seen. The sun, which turned the tops of the Pillars to gold as I entered, crept down inch by inch till it beat upon my head and clothed the rock in a red glory. Still no bird appeared. High above the top of the rocks, in the clear thin air of the mountain, a flock of swallows wheeled and sported, uttering an unfamiliar two-note call; butterflies fluttered irresolute, looking frivolous enough in the presence of the eternal hills; gauzy-winged dragonflies zigzagged to and fro, their intense blue gleaming in the sun. The hour for visitors drew near, and my precious solitude was fast slipping away.

Slowly then I walked up the cañon, looking for my singer. Humming-birds were hovering before the bare rock as before a flower, perhaps sipping the water-drops that here and there trickled down, and large hawks, like mere specks against the blue, were soaring, but no wren could I see. At last I reached the end, with its waterfall fountain. Close within this ceaseless sprinkle, on a narrow ledge that was never dry, was placed—I had almost said grew—a bird's nest; whose, it were needless to ask. One American bird, and one only, chooses perpetual dampness for his environment,—the American dipper, or water ouzel.

Here I paused to muse over the spray-soaked cradle on the rock. In this strange place had lived a bird so eccentric that he prefers not only to nest under a continuous shower, through which he must constantly pass, but to spend most of his life in, not on the water. Shall we call him a fool or a philosopher? Is the water a protection, and from what? Has "damp, moist unpleasantness" no terrors for his fine feathers? Where now were the nestlings whose lullaby had been the music of the falling waters? Down that sheer rock, perhaps into the water at its foot, had been the first flight of the ouzel baby. Why had I come too late to see him?

But the hours were passing, while I had not seen, and, what was worse, had not heard my first charmer, the cañon wren. Leaving these perplexing conundrums unsolved, I turned slowly back down the walk, to resume my search. Perhaps fifty feet from the ouzel nest, as I lingered to admire the picturesque rapids in the brook, a slight movement drew my attention to a little projection on a stone, not six feet from me, where a small chipmunk sat pertly up, holding in his two hands, and eagerly nibbling—was it, could it be a strawberry in this rocky place?

Of course I stopped instantly to look at this pretty sight. I judged him to be a youngster, partly because of his evident fearlessness of his hereditary enemy, a human being; more on account of the saucy way in which he returned my stare; and most, perhaps, from the appearance of absorbing delight, in which there was a suggestion of the unexpected, with which he discussed that sweet morsel. Closely I watched him as he turned the treasure round and round in his deft little paws, and at last dropped the rifled hull. Would he go for another, and where? In an instant, with a parting glance at me, to make sure that I had not moved, he scrambled down his rocky throne, and bounded in great leaps over the path to a crumpled paper, which I saw at once was one of the bags with which tourists sow the earth. But its presence there did not rouse in my furry friend the indignation it excited in me. To him it was a treasure-trove, for into it he disappeared without a moment's hesitation; and almost before I had jumped to the conclusion that it contained the remains of somebody's luncheon, he reappeared, holding in his mouth another strawberry, bounded over the ground to his former seat, and proceeded to dispose of that one, also. The scene was so charming and his pleasure so genuine that I forgave the careless traveler on the spot, and only wished I had a kodak to secure a permanent picture of this unique strawberry festival.

As I loitered along, gazing idly at the brook, ever listening and longing for the wren song, I was suddenly struck motionless by a loud, shrill, and peculiar cry. It was plainly a bird voice, and it seemed to come almost from the stream itself. It ceased in a moment, and then followed a burst of song, liquid as the singing of the brook, and enchantingly sweet, though very low. I was astounded. Who could sing like that up in this narrow mountain gorge, where I supposed the cañon wren was king?

At the point where I stood, a straggling shrub, the only one for rods, hung over the brink. I silently sank to a seat behind it, lest I disturb the singer, and remained without movement. The baffling carol went on for some seconds, and for the only time in my life I wished I could put a spell upon brook-babble, that I might the better hear.

Cautiously I raised my glass to my eyes, and examined the rocks across the water, probably eight feet from me. Then arose again that strange cry, and at the same instant my eye fell upon a tiny ledge, level with the water, and perhaps six inches long, on which stood a small fellow-creature in great excitement. He was engaged in what I should call "curtsying"; that is, bending his leg joint, and dropping his plump little body for a second, then bobbing up to his fullest height, repeating the performance constantly,—looking eagerly out over the water the while, evidently expecting somebody. This was undoubtedly the bird's manner of begging for food,—a very pretty and well-bred way, too, vastly superior to the impetuous calls and demands of some young birds. The movement was "dipping," of course, and he was the dipper, or water ouzel baby, that had been cradled in that fountain-dashed nest by the fall. He was not long out of it, either; for though fully dressed in his modest slate-color, with white feet, and white edgings to many of his feathers, he had hardly a vestige of a tail. He was a winsome baby, for all that.

While I studied the points of the stranger, breathless lest he should disappear before my eyes, he suddenly burst out with the strange call I had heard. It was clearly a cry of joy, of welcome, for out of the water, up on to the ledge beside him, scrambled at that moment a grown-up ouzel. He gave one poke into the wide-open mouth of the infant, then slipped back into the water, dropped down a foot or more, climbed out upon another little shelf in the rock, and in a moment the song arose. I watched the singer closely. The notes were so low and so mingled with the roar of the brook that even then I should not have been certain he was uttering them if I had not seen his throat and mouth distinctly. The song was really exquisite, and as much in harmony with the melody of the stream as the voice of the English sparrow is with the city sounds among which he dwells, and the plaintive refrain of the meadow-lark with the low-lying, silent fields where he spends his days.

But little cared baby ouzel for music, however ravishing. What to his mind was far more important was food,—in short, worms. His pretty begging continued, and the daring notion of attempting a perilous journey over the foot of water that separated him from his papa plainly entered his head. He hurried back and forth on the brink with growing agitation, and was seemingly about to plunge in, when the singer again entered the water, brought up another morsel, and then stood on the ledge beside the eager youngling, "dipping" occasionally himself, and showing every time he winked—as did the little one, also—snowy-white eyelids, in strange contrast to the dark slate-colored plumage.

This aesthetic manner of discharging family duties, alternating food for the body with rapture of the soul, continued for some time, probably until the young bird had as much as was good for him; and then supplies were cut off by the peremptory disappearance of the purveyor, who plunged with the brook over the edge of a rock, and was seen no more.

A little later a grown bird appeared, that I supposed at first was the returning papa, but a few moments' observation convinced me that it was the mother; partly because no song accompanied the work, but more because of the entirely different manners of the newcomer. Filling the crop of that importunate offspring of hers was, with this Quaker-dressed dame, a serious business that left no time for rest or recreation. Two charmed hours I sat absorbed, watching the most wonderful evolutions one could believe possible to a creature in feathers.

At the point where this little drama was enacted, the brook rushed over a line of pebbles stretching from bank to bank, lying at all angles and of all sizes, from six to ten inches in diameter. Then it ran five or six feet quietly, around smooth rocks here and there above the water, and ended by plunging over a mass of bowlders to a lower level. The bird began by mounting one of those slippery rounded stones, and thrusting her head under water up to her shoulders. Holding it there a few seconds, apparently looking for something, she then jumped in where the turmoil was maddest, picked an object from the bottom, and, returning to the ledge, gave it to baby.

The next moment, before I had recovered from my astonishment at this feat of the ouzel, she ran directly up the falls (which, though not high, were exceedingly lively), being half the time entirely under water, and exactly as much at her ease as if no water were there;

though how she could stand in the rapid current, not to speak of walking straight up against it, I could not understand.

Often she threw herself into the stream, and let it carry her down, like a duck, a foot or two, while she looked intently on the bottom, then simply walked up out of it on to a stone. I could see that her plumage was not in the least wet; a drop or two often rested on her back when she came out, but it rolled off in a moment. She never even shook herself. The food she brought to that eager youngling every few minutes looked like minute worms, doubtless some insect larvæ.

Several times this hard-working mother plunged into the brook where it was shallow, ran or walked down it, half under water, and stopped on the very brink of the lower fall, where one would think she could not even stand, much less turn back and run up stream, which she did freely. This looked to me almost as difficult as for a man to stand on the brink of Niagara, with the water roaring and tumbling around him. Now and then the bird ran or flew up, against the current, and entirely under water, so that I could see her only as a dark-colored moving object, and then came out all fresh and dry beside the baby, with a mouthful of food. I should hardly dare to tell this, for fear of raising doubts of my accuracy, if the same thing had not been seen and reported by others before me. Her crowning action was to stand with one foot on each of two stones in the middle and most uproarious part of the little fall, lean far over, and deliberately pick something from a third stone.

All this was no show performance, even no frolic, on the part of the ouzel,—it was simply her every-day manner of providing for the needs of that infant; and when she considered the duty discharged for the time, she took her departure, very probably going at once to the care of a second youngster who awaited her coming in some other niche in the rocks.

Finding himself alone again, and no more dainties coming his way, the young dipper turned for entertainment to the swift-running streamlet. He went down to the edge, stepping easily, never hopping; but when the shallow edge of the water ran over his pretty white toes, he hastily scampered back, as if afraid to venture farther. The clever little rogue was only coquetting, however, for when he did at last plunge in he showed himself very much at home. He easily crossed a turbulent bit of the brook, and when he was carried down a little he scrambled without trouble up on a stone. All the time, too, he was peering about after food; and in fact it was plain that his begging was a mere pretense,—he was perfectly well able to look out for himself. Through the whole of these scenes not one of the birds, old or young, had paid the slightest attention to me, though I was not ten feet from them.

During the time I had been so absorbed in my delightful study of domestic life in the ouzel family, the other interesting resident of the cañon—the elusive cañon wren—had been forgotten. Now, as I noticed that the day was waning, I thought of him again, and, tearing myself away from the enticing picture, leaving the pretty baby to his own amusements, I returned to the famous Pillars, and planted myself before my rock, resolved to stay there till the bird appeared.

No note came to encourage me, but, gazing steadily upward, after a time I noticed something that looked like a fly running along the wall. Bringing my glass to my eyes, I found that it was a bird, and one of the white-throated family I so longed to see. She—for her silence and her ways proclaimed her sex—was running about where appeared to be nothing but perpendicular rock, flirting her tail after the manner of her race, as happy and as unconcerned as if several thousand feet of sheer cliff did not stretch between her and the brook at its foot. Her movements were jerky and wren-like, and every few minutes she flitted into a tiny crevice that seemed, from my point of view, hardly large enough to admit even her minute form. She was dressed like the sweet singer of yesterday, and the door she entered so familiarly was the same I had seen him interested in. I guessed that she was his mate.

The bird seemed to be gathering from the rock something which she constantly carried into the hole. Possibly there were nestlings in that snug and inaccessible home. To discover if my conjectures were true, I redoubled my vigilance, though it was neck-breaking work, for so narrow was the cañon at that point that I could not get far enough away for a more level view.

Sometimes the bustling little wren flew to the top of the wall, about twenty feet above her front door, as it looked to me (it may have been ten times that). Over the edge she instantly disappeared, but in a few minutes returned to her occupation on the rock. Upon the earth beneath her sky parlor she seemed never to turn her eyes, and I began to fear that I should get no nearer view of the shy cliff-dweller.

Finally, however, the caprice seized the tantalizing creature of descending to the level of mortals, and the brook. Suddenly, while I looked, she flung herself off her perch, and fell—down—down—down—disappearing at last behind a clump of weeds at the bottom. Was she killed? Had she been shot by some noiseless air-gun? What had become of the tiny wren? I sprang to my feet, and hurried as near as the intervening stream would allow, when lo! there she was, lively and fussy as ever, running about at the foot of the cliff, searching, searching all the time, ever and anon jumping up and pulling from the rock something that clung to it.

When the industrious bird had filled her beak with material that stuck out on both sides, which I concluded to be some kind of rock moss, she started back. Not up the face of that blank wall, loaded as she was, but by a strange path that she knew well, up which I watched her wending her way to her proper level. This was a cleft between two solid bodies of rock, where, it would seem, the two walls, in settling together for their lifelong union, had broken and crumbled, and formed between them a sort of crack, filled with unattached bowlders, with crevices and passages, sometimes perpendicular, sometimes horizontal. Around and through these was a zigzag road to the top, evidently as familiar to that atom of a bird as Broadway is to some of her fellow-creatures, and more easily traversed, for she had it all to herself.

The wren flew about three feet to the first step of her upward passage, then ran and clambered nearly all the rest of the way, darting behind jutting rocks and coming out the other side, occasionally flying a foot or two; now pausing as if for an observation, jerking her tail upright and letting it drop back, wren-fashion, then starting afresh, and so going on till she reached the level of her nest, when she flew across the (apparent) forty or fifty feet, directly into the crevice. In a minute she came out, and without an instant's pause flung herself down again.

I watched this curious process very closely. The wren seemed to close her wings; certainly she did not use them, nor were they in the least spread that I could detect. She came to the ground as if she were a stone, as quickly and as directly as a stone would have fallen; but just before touching the ground she spread her wings, and alighted lightly on her feet. Then she fell to her labor of collecting what I suppose was nesting material, and in a few minutes started up again by the roundabout road to the top. Two hours or more, with gradually stiffening neck, I spent with the wren, while she worked constantly and silently, and not once during all that time did the singer appear.

What the scattering parties of tourists, who from time to time passed me, thought of a silent personage sitting in the cañon alone, staring intently up at a blank wall of rock, I did not inquire. Perhaps that she was a verse-writer seeking inspiration; more likely, however, a harmless lunatic musing over her own fancies.

I know well what I thought of them, from the glimpses that came to me as I sat there; some climbing over the sharp-edged rocks, in tight boots, delicate kid gloves, and immaculate traveling costumes, and panting for breath in the seven thousand feet altitude; others uncomfortably seated on the backs of the scraggy little burros, one of whom was so interested in my proceedings

that he walked directly up and thrust his long, inquiring
ears into my very face, spite of the resistance of his rider,
forcing me to rise and decline closer acquaintance. One
of the melancholy procession was loaded with a heavy
camera, another equipped with a butterfly net; this one
bent under the weight of a big basket of luncheon, and
that one was burdened with satchels and wraps and um-
brellas. All were laboriously trying to enjoy themselves,
but not one lingered to look at the wonder and the
beauty of the surroundings. I pitied them, one and all,
feeling obliged, as no doubt they did, to "see the sights;"
tramping the lovely cañon to-day, glancing neither to
right nor left; whirling through the Garden of the Gods
to-morrow; painfully climbing the next day the burro
track to the Grave, the sacred point where

"Upon the wind-blown mountain spot
Chosen and loved as best by her,
Watched over by near sun and star,
Encompassed by wide skies, she sleeps."

Alas that one cannot quote with truth the remaining
lines!

"And not one jarring murmur creeps
Up from the plain her rest to mar."

For now, at the end of the toilsome passage, that place
which should be sacred to loving memories and tender
thoughts, is desecrated by placards and picnickers, de-
faced by advertisements, strewn with the wrapping-
paper, tin cans, and bottles with which the modern
globe-trotter marks his path through the beautiful and
sacred scenes in nature.[1]

In this uncomfortable way the majority of summer tourists spend day after day, and week after week; going home tired out, with no new idea gained, but happy to be able to say they have been here and there, beheld this cañon, dined on that mountain, drank champagne in such a pass, and struggled for breath on top of "the Peak." Their eyes may indeed have passed over these scenes, but they have not seen one thing.

Far wiser is he (and more especially she) who seeks out a corner obscure enough to escape the eyes of the "procession," settles himself in it, and spends fruitful and delightful days alone with nature; never hasting nor rushing; seeing and studying the wonders at hand, but avoiding "parties" and "excursions;" valuing more a thorough knowledge of one cañon than a glimpse of fifty; caring more to appreciate the beauties of one mountain than to scramble over a whole range; getting into such perfect harmony with nature that it is as if he had come into possession of a new life; and from such an experience returning to his home refreshed and invigorated in mind and body.

Such were my reflections as the sun went down, and I felt, as I passed out through the gate, that I ought to double my entrance fee, so much had my life been enriched by that perfect day alone in Cheyenne Cañon.

[1] Since the above was written, I am glad to learn that, because of this vandalism, the remains of "H. H." have been removed to the cemetery at Colorado Springs.

IN THE MIDDLE COUNTRY

For all the woods are shrill with stress of song,
Where soft wings flutter down to new-built nests,
And turbulent sweet sounds are heard day long,
As of innumerable marriage feasts.

Charles Lotin Hildreth

VIII. AT FOUR O'CLOCK
IN THE MORNING

Four o'clock in the morning is the magical hour of the day. I do not offer this sentiment as original, nor have I the slightest hope of converting any one to my opinion; I merely state the fact.

For years I had known it perfectly well; and fortified by my knowledge, and bristling with good resolutions, I went out every June determined to rise at that unnatural hour. Nothing is easier than to get up at four o'clock—the night before; but when morning comes, the point of view is changed, and all the arguments that arise in the mind are on the other side; sleep is the one thing desirable. The case appeared hopeless. Appeals from Philip drunk (with sleep) to Philip sober did not seem to avail; for whatever the latter decreed, the former would surely disobey.

But last June I found my spur; last summer I learned to get up with eagerness, and stay up with delight. This was effected by means of an alarm, set by the evening's wakefulness, that had no mercy on the morning's sleepiness. The secret is—a present interest. What may be going on somewhere out of sight and hearing in the world is a matter of perfect indifference; what is heard and seen at the moment is an argument that no one can resist.

I got my hint by the accident of some shelled corn being left on the ground before my window, and so attracting a four o'clock party, consisting of blackbirds, blue jays, and doves. I noticed the corn, but did not think of the pleasure it would give me, until the next morning, when I was awakened about four o'clock by loud and excited talk in blackbird tones, and hurried to the window, to find that I had half the birds of the neighborhood before me.

Most in number, and most noisy, were the common blackbirds, who just at that time were feeding their young in a grove of evergreens back of the house, where they had set up their nurseries in a crowd, as is their custom. It is impossible to take this bird seriously, he is so irresistibly ludicrous. His manners always suggest to me a peculiar drollery; of a person of the old-fashioned sort, as we read of him, and I promised myself some amusement from the study of him at short range; I was not disappointed.

My greeting as I took my seat at the open window, unfortunately without blinds to screen me, was most comical. A big pompous fellow turned his wicked-looking white eye upon me, he drew himself into a queer humped-up position, with all his feathers on end, and apparently by a strong effort squeezed out a husky and squeaky, yet loud cry of two notes, which sounded exactly like "Squee-gee!"

I was so astounded that I laughed in his face; at which he repeated it with added emphasis, then turned his back on me, as unworthy of notice away up in my window, and gave his undivided attention to a specially large grain of corn which had been unearthed by a meek-looking neighbor, and appropriated by him, in the most lordly manner. His bearing at the moment was superb and stately in a degree of which only a bird who walks is capable; one cannot be dignified who is obliged to hop.

I thought his greeting was a personal one to show contempt—which it did emphatically—to the human race in general, and to me in particular, but I found later that it was the ordinary blackbird way of being offensive; it was equivalent to "Get out!" or "Shut up!" or some other of the curt and rude expressions in use by bigger folk than blackbirds.

If a bird alighted too near one of these arrogant fellows on the ground, he was met with the same expletive, and if he was about the same size he "talked back." The number and variety of utterances at their command was astonishing; I was always being surprised with a new one. Now a blackbird would fly across the lawn, making a noise exactly like a boy's tin trumpet, and repeating it as long as he was within hearing, regarding it, seemingly, as an exceptionally great feat. Again one would seize a kernel of corn, burst out with a convulsive cry, as if he were choking to death, and fly off with his prize, in imminent danger of his life, as I could not but feel.

The second morning a youngster came with his papa to the feast, and he was droller, if possible, than his elders. He followed his parent around, with head lowered and mouth wide open, fairly bawling in a loud yet husky tone.

The young blackbird does not appear in the glossy suit of his parents. His coat is rusty in hue, and his eye is dark, as is proper in youth. He is not at all backward in speaking his mind, and his sole desire at this period of his life being food, he demands it with an energy and persistence that usually insures success.

In making close acquaintance with them, one cannot help longing to prescribe to the whole blackbird family something to clear their bronchial tubes; every tone is husky, and the student involuntarily clears his own throat as he listens.

I was surprised to find the blackbirds so beautiful. When the sun was near setting, and struck across the grass its level rays, they were really exquisite; their heads a brilliant metallic blue, and all back of that rich bronze or purple, all over as glossy as satin. The little dames are somewhat smaller, and a shade less finely dressed than their bumptious mates; but that does not make them meek—far from it! and they are not behind their partners in eccentric freaks. Sometimes one would apparently attempt a joke by starting to fly, and passing so near the head of one of the dignitaries on the ground that he would involuntarily start and "duck" ingloriously. On one occasion a pair were working peaceably together at the corn, when she flirted a bit of dirt so that it flew toward him. He dashed furiously at her. She gave one hop which took her about a foot away, and then it appeared that she coveted a kernel of corn that was near him when the offense was given, for she instantly jumped back and pounced upon it as if she expected to be annihilated. He ran after her and drove her off, but she kept her prize.

Eating one of those hard grains was no joke to anybody without teeth, and it was a serious affair to one of the blackbirds. He took it into his beak, dropped both head and tail, and gave his mind to the cracking of the sweet morsel. At this time he particularly disliked to be disturbed, and the only time I saw one rude to a youngster was when struggling with this difficulty. While feeding the nestlings, they broke the kernels into bits, picked up all the pieces, filling the beak the whole length, and then flew off with them.

But they were not always allowed to keep the whole kernel. They were generally attended while on the ground by a little party of thieves, ready and waiting to snatch any morsel that was dropped. These were, of course, the English sparrows. They could not break corn, but they liked it for all that, so they used their wits to secure it, and of sharpness these street birds have no lack. The moment a blackbird alighted on the grass, a sparrow or two came down beside him, and lingered around, watching eagerly. Whenever a crumb dropped, one rushed in and snatched it, and instantly flew from the wrath to come.

The sparrows had not been at this long before some of the wise blackbirds saw through it, and resented

it with proper spirit. One of them would turn savagely
after the sparrow who followed him, and the knowing
rascal always took his departure. It was amusing to see a
blackbird working seriously on a grain, all his faculties
absorbed in the solemn question whether he should succeed in cracking his nut, while two or three feathered pil-
ferers stood as near as they dared, anxiously waiting till
the great work should be accomplished, the hard shell
should yield, and some bits should fall.

About five days after the feast was spread, the
young came out in force, often two of them following
one adult about on the grass, running after him so
closely that he could hardly get a chance to break up the
kernel; indeed, he often had to fly to a tree to prepare the
mouthfuls for them. The young blackbird has not the
slightest repose of manner; nor, for that matter, has the
old one either. The grown-ups treated the young well, almost always; they never "squee-gee'd" at them, never
touched them in any way, notwithstanding they were so
insistent in begging that they would chase an adult bird
across the grass, calling madly all the time, and fairly
force him to fly away to get rid of them.

Once two young ones got possession of the only
spot where corn was left, and so tormented their elders
who came that they had to dash in and snatch a kernel
when they wanted one. One of the old ones danced
around these two babies in a little circle a foot in diameter, the infants turning as he moved, and ever present-
ing open beaks to him. It was one of the funniest exhibi-
tions I ever saw. After going around half a dozen times,
the baffled blackbird flew away without a taste.

When the two had driven every one else off the ground by their importunities, one of them plucked up spirit to try managing the corn for himself. Like a little man he stopped bawling, and began exercising his strength on the sweet grain. Upon this his neighbor, instead of following his example, began to beg of him! fluttering his wings, putting up his beak, and almost pulling the corn out of the mouth of the poor little fellow struggling with his first kernel!

Sometimes a young one drove his parent all over a tree with his supplications. Higher and higher would go the persecuted, with his tormentor scrambling, and half flying after, till the elder absolutely flew away, much put out.

Long before this time the corn had been used up. But I could not bear to lose my morning entertainment, for all these things took place between four and six A. M. —so I made a trip to the village, and bought a bag of the much desired dainty, some handfuls of which I scattered every night after birds were abed, ready for the sunrise show. Blackbirds were not the only guests at the feast; there were the doves,—mourning, or wood-doves,—who dropped to the grass, serene as a summer morning, walking around in their small red boots, with mincing steps and fussy little bows. Blue jays, too, came in plenty, selected each his grain and flew away with it. Robins, seeing all the excitement, came over from their regular hunting-ground, but never finding anything so attractive as worms, they soon left.

The corn feast wound up with a droll excitement. One day a child from the house took her doll out in the grass to play, set it up against a tree trunk, and left it there. It had long light hair which stood out around the head, and it did look rather uncanny, but it was amusing to see the consternation it caused. Blue jays came to trees near by, and talked in low tones to each other; then one after another swooped down toward it; then they all squawked at it, and finding this of no avail, they left in a body.

The robins approached cautiously, two of them, calling constantly, "he! he! he!" One was determined not to be afraid, and came nearer and nearer, till within about a foot of the strange object and behind it, when suddenly he started as though shot, jumped back, and both flew in a panic.

Soon after this a red-headed woodpecker alighted on the trunk of the elm, preparatory to helping himself to a grain of corn. The moment his eyes fell upon madam of the fluffy hair, he burst out with a loud, rapid wood-pecker "chitter," gradually growing higher in key and louder in tone. The blue jay flew down from the nest across the yard, and another came from behind the house; both perched near and stared at him, and then be-gan to talk in low tones. A robin came hastily over and gazed at the usually silent red-head, and apparently it was to all as strange a performance as it was to me, or possibly they recognized that it was a cry of warning against danger.

After he had us all aroused, the bird suddenly fell to silence, and resumed his ordinary manner, but he did not go after corn. I suppose the harangue was addressed to the doll.

That was the last scene in the first act of the corn feast, for the blackbirds had become so numerous and so noisy that they made morning hideous to the whole household, and I stopped the supplies for several days, till these birds ceased to expect anything, and so came no more, and then I spread a fresh breakfast-table for more interesting guests, whose manners and customs I studied for weeks.

I was invariably startled wide awake on these mornings by a bird note, and sprang up, to see at one glance that

"Day had awakened all things that be,
The lark and the thrush, and the swallow free,"

and that my party was already assembled; one or two cardinals—or redbirds, as they are often called—on the grass, with the usual attendance of English sparrows, and the red-headed woodpecker in the elm, surveying the lawn, and considering which of the trespassers he should fall upon. It was the work of one minute to get into my wraps and seat myself, with opera glass, at the wide-open window.

My first discovery made, however, during the blackbird reign, was that four o'clock is the most lovely part of the day. All the dust of human affairs having settled during the hours of sleep, the air is fresh and sweet, as if just made; and generally, just before sunrise, the foliage is at perfect rest,—the repose of night still lingering, the world of nature as well as of men still sleeping.

The first thing one naturally looks for, as birds begin to waken, is a morning chorus of song. True bird-lovers, indeed, long for it with a longing that cannot be told. But alas, every year the chorus is withdrawing more and more to the woods, every year it is harder to find a place where English sparrows are not in possession; and it is one of the most grievous sins of that bird that he spoils the song, even when he does not succeed in driving out the singer. A running accompaniment of harsh and interminable squawks overpowers the music of meadow-lark and robin, and the glorious song of the thrush is fairly murdered by it. One could almost forgive the sparrow his other crimes, if he would only lie abed in the morning; if he would occasionally listen, and not forever break the peace of the opening day with his vulgar brawling. But the subject of English sparrows is maddening to a lover of native birds; let us not defile the magic hour by considering it.

The most obvious resident of the neighborhood, at four o'clock in the morning, was always the golden-winged woodpecker, or Northern flicker. Though he scorned the breakfast I offered, having no vegetarian proclivities, he did not refuse me his presence. I found him a character, and an amusing study, and I never saw his tribe so numerous and so much at home.

Though largest in size of my four o'clock birds, and most fully represented (always excepting the English sparrows), the golden-wing was not in command. The autocrat of the hour, the reigning power, was quite a different personage, although belonging to the woodpecker family. It was a red-headed woodpecker who assumed to own the lawn and be master of the feast. This individual was marked by a defect in plumage, and had been a regular caller since the morning of my arrival. During the blackbird supremacy over the corn supply he had been hardly more than a spectator, coming to the trunk of the elm and surveying the assembly of blue jays, doves, blackbirds, and sparrows with interest, as one looks down upon a herd with whom he has nothing in common. But when those birds departed, and the visitors were of a different character, mostly cardinals, with an occasional blue jay, he at once took the place he felt belonged to him—that of dictator.

The Virginia cardinal, a genuine F. F. V., and a regular attendant at my corn breakfast, was a subject of special study with me; indeed, it was largely on his account that I had set up my tent in that part of the world. I had all my life known him as a tenant of cages, and it struck me at first as very odd to see him flying about freely, like other wild birds. No one, it seemed to me, ever looked so out of place as this fellow of elegant manners, aristocratic crest, and brilliant dress, hopping about on the ground with his exaggerated little hops, tail held stiffly up out of harm's way, and uttering sharp "tsips." One could not help the feeling that he was altogether too fine for this common work-a-day existence; that he was intended for show; and that a gilded cage was his proper abiding-place, with a retinue of human servants to minister to his comfort. Yet he was modest and unassuming, and appeared really to enjoy his life of hard work; varying his struggles with a kernel of hard corn on the ground, where his color shone out like a flower against the green, with a rest on a spruce-tree, where

"Like a living jewel he sits and sings;"

and when he had finished his frugal meal, departing, if nothing hurried him, with a graceful, loitering flight, in

which each wing-beat seemed to carry him but a few inches forward, and leave his body poised, an infinitesimal second for another beat. With much noise of fluttering wings he would start for some point, but appear not to care much whether he got there. He was never in haste unless there was something to hurry him, in which he differed greatly from some of the fidgety, restless personages I have known among the feathered folk.

The woodpecker's way of making himself disagreeable to this distinguished guest, was to keep watch from his tree (an elm overlooking the supply of corn) till he came to eat, and then fly down, aiming for exactly the spot occupied by the bird on the ground. No one, however brave, could help "getting out from under," when he saw this tricolored whirlwind descending upon him. The cardinal always jumped aside, then drew himself up, crest erect, tail held at an angle of forty-five degrees, and faced the woodpecker, calm, but prepared to stand up for his right to life, liberty, and the pursuit of his breakfast. Sometimes they had a little set-to, with beaks not more than three inches apart, the woodpecker making feints of rushing upon his vis-à-vis, and the cardinal jumping up ready to clinch, if a fight became necessary. It never went quite so far as that, though they glared at each other, and the cardinal uttered a little whispered "ha!" every time he sprang up.

The Virginian's deliberate manner of eating made peace important to him. He took a grain of hard corn in his mouth, lengthwise; then working his sharp-edged beak, he soon succeeded in cutting the shell of the kernel through its whole length. From this he went on turning it with his tongue, and still cutting with his beak, till the whole shell rolled out of the side of his mouth in one long piece, completely cleared from its savory contents.

The red-head, on the contrary, took his grain of corn to a branch, or sometimes to the trunk of a tree, where he sought a suitable crevice in the bark or in a crotch, placed his kernel, hammered it well in till firm and safe, and then proceeded to pick off pieces and eat them daintily, one by one. Sometimes he left a kernel there, and I saw how firmly it was wedged in, when the English sparrow discovered his store, fell upon it, and dug it out. It was a good deal of work for a strong-billed, persistent sparrow to dislodge a grain thus placed. But of course he never gave up till he could carry it off, probably because he saw that some one valued it; for since he was unable to crack a grain that was whole, it must have been useless to him. Sometimes the woodpecker wedged the kernel into a crevice in the bark of the trunk, then broke it up, and packed the pieces away in other niches; and I have seen an English sparrow go carefully over the trunk, picking out and eating these tidbits. That, or something else, has taught sparrows to climb tree trunks, which they do, in the neighborhood I speak of, with as much ease as a woodpecker. I have repeatedly seen them go the whole length of a tall elm trunk; proceeding by little hops, aided by the wings, and using the tail for support almost as handily as a woodpecker himself.

The red-head's assumption of being monarch of all he surveyed did not end with the breakfast-table; he seemed to consider himself guardian and protector of the whole place. One evening I was drawn far down on the lawn by a peculiar cry of his. It began with a singular performance which I have already described, a loud, rapid "chit-it-it-it-it," increasing in volume and rising in pitch, as though he were working himself up to some deed of desperation. In a few minutes, however, he appeared to get his feelings under control, and dropped to a single-note cry, often repeated. It differed widely from his loud call, "wok! wok! wok!" still more from the husky tones of his conversation with others of his kind; neither was it like the war-cries with which he intimated to another bird that he was not invited to breakfast. I thought there must be trouble brewing, especially as mingled with it was an occasional excited "pe-auk!" of a flicker. When I reached the spot, I found a curious party, consisting of two doves and three flickers, assembled on one small tree, with the woodpecker on an upper branch, as though addressing his remarks to them.

As I drew near the scene of the excitement, the doves flew, and then the golden-wings; but the red-head held his ground, though he stopped his cries when he saw help coming. In vain I looked about for the cause of the row; everything was serene. It was a beautiful quiet evening, and not a child, nor a dog, nor anything in sight to make trouble. The tree stood quite by itself, in the midst of grass that knew not the clatter of the lawn-mower.

I stood still and waited; and I had my reward, for after a few minutes' silence I saw a pair of ears, and then a head, cautiously lifted above the grass, about fifteen feet from the tree. The mystery was solved; it was a cat, whom all birds know as a creature who will bear watching when prowling around the haunts of bird families. I am fond of pussy, but I deprecate her taste for game, as I do that of some other hunters, wiser if not better than she. I invited her to leave this place, where she plainly was unwelcome, by an emphatic "scat!" and a stick tossed her way. She instantly dropped into the grass and was lost to view; and as the woodpecker, whose eyes were sharper and his position better than mine, said no more, I concluded she had taken the hint and departed.

IX. THE LITTLE REDBIRDS

When the little redbirds began to visit the lawn there were exciting times. At first they ventured only to the trees overlooking it; and the gayly dressed father who had them in charge reminded me of nothing so much as a fussy young mother. He was alert to the tips of his toes, and excited, as if the whole world was thirsting for the life of those frowzy-headed youngsters in the maple. His manner intimated that nobody ever had birdlings before; indeed, that there never had been, or could be, just such a production as that young family behind the leaves. While they were there, he flirted his tail, jerked himself around, crest standing sharply up, and in every way showed his sense of importance and responsibility.

As for the young ones, after they had been hopping about the branches a week or so, and papa had grown less madly anxious if one looked at them, they appeared bright and spirited, dressed in the subdued and tasteful hues of their mother, with pert little crests and dark beaks. They were not allowed on the grass, and they waited patiently on the tree while their provider shelled a kernel and took it up to them. The cardinal baby I found to be a self-respecting individual, who generally waits in patience his parents' pleasure, though he is not too often fed. He is not bumptious nor self-assertive, like many others; he rarely teases, and is certainly altogether a well-mannered and proper young person. After a while, as the youngsters learned strength and speed on the wing, they came to the table with the grown-ups, and then I saw there were three spruce young redbirds, all under the care of their gorgeous papa.

No sooner did they appear on the ground than trouble began with the English-sparrow tribe. The grievance of these birds was that they could not manage the tough kernels. They were just as hungry as anybody, and just as well-disposed toward corn, but they had not sufficient strength of beak to break it. They did not, however, go without corn, for all that. Their game was the not uncommon one of availing themselves of the labor of others; they invited themselves to everybody's breakfast-table, though, to be sure, they had to watch their chances in order to secure a morsel, and escape the wrath of the owner thereof.

The cardinal was at first a specially easy victim to this plot. He took the whole matter most solemnly, and was so absorbed in the work, that if a bit dropped, in the process of separating it from the shell, as often happened, he did not concern himself about it till he had finished what he had in his mouth, and then he turned one great eye on the ground, for the fragments which had long before been snatched by sparrows and gone down sparrow throats. The surprise and the solemn stare with which he "could hardly believe his eyes" were exceedingly droll. After a while he saw through their little game, and took to watching, and when a sparrow appeared too much interested in his operations, he made a feint of going for him, which warned the gamin that he would better look out for himself.

It did not take these sharp fellows long to discover that the young redbird was the easier prey, and soon every youngster on the ground was attended by a sparrow or two, ready to seize upon any fragment that fell. The parent's way of feeding was to shell a kernel and then give it to one of the little ones, who broke it up and ate it. From waiting for fallen bits, the sparrows, never being repulsed, grew bolder, and finally went so far as actually to snatch the corn out of the young cardinals' beaks. Again and again did I see this performance: a sparrow grab and run (or fly), leaving the baby astonished and dazed, looking as if he did not know exactly what had happened, but sure he was in some way bereaved.

One day, while the cardinal family were eating on the grass, the mother of the brood came to a tree near by. At once her gallant spouse flew up there and offered her the mouthful he had just prepared, then returned to his duties. She was rarely seen on the lawn, and I judged that she was sitting again.

Sometimes, when the youngsters were alone on the ground, I heard a little snatch of song, two or three notes, a musical word or two of very sweet quality. The woodpecker, autocrat though he assumed to be, did not at first interfere with the young birds; but as they became more and more independent and grown up, he began to consider them fair game, and to come down on them with a rush that scattered them; not far, however; they were brave little fellows.

At last, after four weeks of close attention, the cardinal made up his mind that his young folk were babies no longer, and that they were able to feed themselves. I was interested to see his manner of intimating to his young hopefuls that they had reached their majority. When one begged of him, in his gentle way, the parent turned suddenly and gave him a slight push. The urchin understood, and moved a little farther off; but perhaps the next time he asked he would be fed. The young bird learned the lesson, however, and in less than two days from the first hint they became almost entirely independent.

One morning the whole family happened to meet at table. The mother came first, and then the three young ones, all of whom were trying their best to feed themselves. At last came their "natural provider;" and one of the juveniles, who found the grains almost unmanageable, could not help begging of him. He gently but firmly drove the pleader away, as if he said, "My son, you are big enough to feed yourself." The little one turned, but did not go; he stood with his back toward his parent, and wings still fluttering. Then papa flew to a low branch of the spruce-tree, and instantly the infant followed him, still begging with quivering wings. Suddenly the elder turned, and I expected to see him annihilate that beggar, but, to my surprise, he fed him! He could not hold out against him! He had been playing the stern parent, but could not keep it up. It was a very pretty and very human-looking performance.

A day or two after the family had learned to take care of themselves, the original pair, the parents of the pretty brood, came and went together to the field, while the younglings appeared sometimes in a little flock, and sometimes one alone; and from that time they were to be rated as grown-up and educated cardinals. A brighter or prettier trio I have not seen. I am almost positive there was but one family of cardinals on the place; and if I am right, those youngsters had been four weeks out of the nest before they took charge of their own food supply. From what I have seen in the case of other young birds, I have no doubt that is the fact.

X. THE CARDINAL'S NEST

While I had been studying four o'clock manners, grave and gay, other things had happened. Most delightful, perhaps, was my acquaintance with a cardinal family at home. From the first I had looked for a nest, and had suffered two or three disappointments. One pair flaunted their intentions by appearing on a tree before my window, "tsipping" with all their might; she with her beak full of hay from the lawn below; he, eager and devoted, assisting by his presence. The important and consequential manner of a bird with building material in mouth is amusing. She has no doubt that what she is about to do is the very most momentous fact in the "Sublime Now" (as some college youth has it). Of course I dropped everything and tried to follow the pair, at a distance great enough not to disturb them, yet to keep in sight at least the direction they took, for they are shy birds, and do not like to be spied upon. But I could not have gauged my distance properly; for, though I thought

I knew the exact cedar-tree she had chosen, I found, to my dismay and regret afterward, that no sign of a nest was there, or thereabout.

Another pair went farther, and held out even more delusive hopes; they actually built a nest in a neighbor's yard, the family in the house maintaining an appearance of the utmost indifference, so as not to alarm the birds till they were committed to that nest. For so little does madam regard the labor of building, and so fickle is she in her fancies, that she thinks nothing of preparing at least two nests before she settles on one. The nest was made on a big branch of cedar, perhaps seven feet from the ground,—a rough affair, as this bird always makes. In it she even placed an egg, and then, for some undiscovered reason, it was abandoned, and they took their domestic joys and sorrows elsewhere.

But now, at last, word came to me of an occupied nest to be seen at a certain house, and I started at once for it. It was up a shady country lane, with a meadow-lark field on one side, and a bobolink meadow on the other. The lark mounted the fence, and delivered his strange sputtering cry,—the first I had ever heard from him (or her, for I believe this is the female's utterance). But the dear little bobolink soared around my head, and let fall his happy trills; then suddenly, as Lowell delightfully pictures him,—

"Remembering duty, in mid-quaver stops,
Just ere he sweeps o'er rapture's tremulous brink,
And 'twixt the winrows most demurely drops,
A decorous bird of business, who provides
For his brown mate and fledglings six besides,
And looks from right to left, a farmer mid his crops."

Nothing less attractive than a cardinal family could draw me away from these rival allurements, but I went on.

The cardinal's bower was the prettiest of the summer, built in a climbing rose which ran riot over a trellis beside a kitchen door. The vine was loaded with buds just beginning to unfold their green wraps to flood the place with beauty and fragrance, and the nest was so carefully tucked away behind the leaves that it could not be seen from the front. Whether from confidence in the two or three residents of the cottage, or because the house was alone so many hours of the day,—the occupants being students, and absent most of the time,—the birds had taken no account of a window which opened almost behind them. From that window one could look into, and touch, if he desired, the little family. But no one who lived there did desire (though I wish to record that one was a boy of twelve or fourteen, who had been taught respect for the lives even of birds), and these birds became so accustomed to their human observers that they paid no attention to them.

The female cardinal is so dainty in looks and manner, so delicate in all her ways, that one naturally expects her to build at least a neat and comely nest, and I was surprised to see a rough-looking affair, similar to the one already mentioned. This might be, in her case, because it was the third nest she had built that summer. One had been used for the first brood. The second had been seized and appropriated to their own use by another pair of birds. (As this was told me, and I cannot vouch for it, I shall not name the alleged thief.) This, the third, was made of twigs and fibres of bark,—or what looked like that,—and was strongly stayed to the rose stems, the largest of which was not bigger than my little finger, and most of them much smaller.

On my second visit I was invited into the kitchen to see the family in the rosebush. It appeared that this was "coming-off" day, and one little cardinal had already taken his fate in his hands when I arrived, soon after breakfast. He had progressed on the journey of life about one foot; and a mere dot of a fellow he looked beside his parents, with a downy fuzz on his head, which surrounded it like a halo, and no sign of a crest. The three nestlings still at home were very restless, crowding, and almost pushing each other out. They could well spare their elder brother, for before he left he had walked all over them at his pleasure; and how he could help it in those close quarters I do not see.

While I looked on, papa came with provisions. At one time the food consisted of green worms about twice as large as a common knitting needle. Three or four of them he held crosswise of his beak, and gave one to each nestling. The next course was a big white grub, which he did not divide, but gave to one, who had considerable difficulty in swallowing it.

I said the birds did not notice the family, but they very quickly recognized me as a stranger. They stood and glared at me in the cardinal way, and uttered some sharp remonstrance; but business was pressing, and I was unobtrusive, so they concluded to ignore me.

The advent of the first redbird baby seemed to give much pleasure, for the head of the family sang a good deal in the intervals of feeding; and both of the pair appeared very happy over it, often alighting beside the wanderer, evidently to encourage him, for they did not always feed. The youngster, after an hour, perhaps, flew about ten feet to a peach-tree, where he struggled violently, and nearly fell before he secured a hold on a twig. Both parents flew to his assistance, but he did not fall, and soon after he flew to a grape trellis, and, with a little clambering, to a stem of the vine, where he seemed pleased to stay,—perhaps because this overlooked the garden whence came all his food.

I stayed two or three hours with the little family, and then left them; and when I appeared the next morning all were gone from the nest. I heard the gentle cries of young redbirds all around, but did not try to look them up, both because I did not want to worry the parents, and because I had already made acquaintance with young cardinals in my four o'clock studies.

The place this discerning pair of birds had selected in which to establish themselves was one of the most charming nooks in the vicinity. Kept free from English sparrows (by persistently destroying their nests), and having but a small and quiet family, it was the delight of cardinals and catbirds. Without taking pains to look for them, one might see the nests of two catbirds, two wood doves, a robin or two, and others; and there were beside, thickets, the delight of many birds, and a row of spruces so close that a whole flock might have nested there in security. In that spot "the quaintly discontinuous lays" of the catbird were in perfection; one song especially was the best I ever heard, being louder and more clear than catbirds usually sing.

As I turned to leave the grounds, the relieved parent, who had not relished my interest in his little folk, mounted a branch, and,

> "Like a pomegranate flower
> In the dark foliage of the cedar-tree,
> Shone out and sang for me."

And thus I left him.

XI. LITTLE BOY BLUE

"The crested blue jay flitting swift."

To know the little boy blue in his domestic life had been my desire for years. In vain did I search far and wide for a nest, till it began to look almost as if the bird intentionally avoided me. I went to New England, and blue jays disappeared as if by magic; I turned my steps to the Rocky Mountains, and the whole tribe betook itself to the inaccessible hills. In despair I abandoned the search, and set up my tent in the middle country, without a thought of the bonny blue bird. One June morning I seated myself by my window, which looked out upon a goodly stretch of lawn dotted with trees of many kinds, and behold the long-desired object right before my eyes!

The blue jay himself pointed it out to me; unconsciously, however, for he did not notice me in my distant window. From the ground, where I was looking at him, he flew directly to a pine-tree about thirty feet high, and there, near the top, sat his mate on her nest. He leaned over her tenderly; she fluttered her wings and opened her mouth, and he dropped into it the tidbit he had brought. Then she stepped to a branch on one side, and he proceeded to attend to the wants of the young family, too small as yet to appear above the edge.

The pine-tree, which from this moment became of absorbing interest, was so far from my window that the birds never thought of me as an observer, and yet so near that with my glass I could see them perfectly. It was also exactly before a thick-foliaged maple, that formed a background against which I could watch the life of the nest, wherever the sunlight fell, and whatever the condition of the sky; so happily was placed my blue jay household.

I observed at once that the jay was very gallant and attentive to his spouse. The first mouthful was for her, even when babies grew clamorous, and she took her share of the work of feeding. Nor did he omit this little politeness when they went to the nest together, both presumably with food for the nestlings. She was a devoted mother, brooding her bantlings for hours every day, till they were so big that it was hard to crowd them back into the cradle; and he was an equally faithful father, working from four o'clock in the morning till after dusk, a good deal of the time feeding the whole family. I acquired a new respect for Cyanocitta cristata.

I had not watched the blue jays long before I was struck with the peculiar character of the feathered world about me, the strange absence of small birds. The neighbors were blackbirds (common grackles), Carolina doves, Northern flickers and red-headed woodpeckers, robins and cardinal grosbeaks, and of course English sparrows, —all large birds, able to hold their own by force of arms, as it were, except the foreigner, who maintained his position by impudence and union, a mob being his weapon of offense and defense. Beside him no small bird lived in the vicinity. No vireo hung there her dainty cup, while her mate preached his interminable sermons from the trees about; no phœbe shouted his woes to an unsympathizing world; no sweet-voiced goldfinch poured out his joyous soul; not a song-sparrow tuned his little lay within our borders. Unseen of men, but no doubt sharply defined to clearer senses than ours, was a line barring them out.

Who was responsible for this state of things? Could it be the one pair of jays in the pine, or the colony of blackbirds the other side of the house? Should we characterize it as a blue jay neighborhood or a blackbird neighborhood? The place was well policed, certainly; robins and blue jays united in that work, though their relations with each other bore the character of an armed neutrality, always ready for a few hot words and a little bluster, but never really coming to blows. We never had the pleasure of seeing a stranger among us. We might hear him approaching, nearer and nearer, till, just as the eager listener fancied he might alight in sight, there would burst upon the air the screech of a jay or the war-cry of a robin, accompanied by the precipitate flight of the whole clan, and away would go the stranger in a most sensational manner, followed by outcries and clamor enough to drive off an army of feathered brigands. This neighborhood, if the accounts of his character are to be credited, should be the congenial home of the kingbird,

—tyrant flycatcher he is named; but as a matter of fact, not only were the smaller flycatchers conspicuous by their absence, but the king himself was never seen, and the flying tribes of the insect world, so far as dull-eyed mortals could see, grew and flourished.

Close scrutiny of every movement of wings, however, revealed one thing, namely, that any small bird who appeared within our precincts was instantly, without hesitation, and equally without unusual noise or special publicity, driven out by the English sparrow; and I became convinced that he, and he alone, was responsible for the presence of none but large birds, who could defy him.

One of the prettiest sights about the pine-tree homestead was the way the jay went up to it. He never imitated the easy style of his mate, who simply flew to a branch below the three that held her treasure, and hopped up the last step. Not he; not so would his knightly soul mount to the castle of his sweetheart and his babies. He alighted much lower, often at the foot of the tree, and passed jauntily up the winding way that led to them, hopping from branch to branch, pausing on each, and circling the trunk as he went; now showing his trim violet-blue coat, now his demure Quaker-drab vest and black necklace; and so he ascended his spiral stair.

There is nothing demure about the blue jay, let me hasten to say, except his vest; there is no pretension about him. He does not go around with the meek manners of the dove, and then let his angry passions rise, in spite of his reputation, as does that "meek and gentle" fellow-creature on occasion. The blue jay takes his life with the utmost seriousness, however it may strike a looker-on. While his helpmeet is on the nest, it is, according to the blue jay code, his duty, as well as it is plainly his pleasure, to provide her with food, which consequently he does; later, it is his province not only to feed, but to protect the family, which also he accomplishes with much noise and bluster. Before the young are out comes his hardest task, keeping the secret of the nest, which obliges him to control his naturally boisterous tendencies; but even in this he is successful, as I saw in the case of a bird whose mate was sitting in an apple-tree close beside a house. There, he was the soul of discretion, and so subdued in manner that one might be in the vicinity all day and never suspect the presence of either. All the comings and goings took place in silence, over the top of the tree, and I have watched the nest an hour at a time without being able to see a sign of its occupancy, except the one thing a sitting bird cannot hide, the tail. And, by the way, how providential—from the bird student's point of view—that birds have tails! They can, it is true, be narrowed to the width of one feather and laid against a convenient twig, but they cannot be wholly suppressed, nor drawn down out of sight into the nest with the rest of the body.

When the young blue jays begin to speak for themselves, and their vigilant protector feels that the precious secret can no longer be kept, then he arouses the neighborhood with the announcement that here is a nest he is bound to protect with his life; that he is engaged in performing his most solemn duty, and will not be disturbed. His air is that so familiar in bigger folk, of daring the whole world to "knock a chip off his shoulder," and he goes about with an appearance of important business on hand very droll to see.

The bearing of the mother of the pine-tree brood was somewhat different from that of her mate, and by their manners only could the pair be distinguished. Whatever may be Nature's reason for dressing the sexes unlike each other in the feathered world,—which I will leave for the wise heads to settle,—it is certainly an immense advantage to the looker-on in birddom. When a pair are facsimiles of each other, as are the jays, it requires the closest observation to tell them apart; indeed, unless there is some defect in plumage, which is not uncommon, it is necessary to penetrate their personal characteristics, to become familiar with their idiosyncrasies of habit and manner. In the pine-tree family, the mother had neither the presence of mind nor the bluster of the partner of her joys. When I came too near the nest tree, she greeted me with a plaintive cry, a sort of "craw! craw!" at the same time "jouncing" herself violently, thus protesting against my intrusion; while he saluted me with squawks that made the welkin ring. Neither of them paid any attention to me, so long as I remained upon a stationary bench not far from their tree; they were used to seeing people in that place, and did not mind them. It was the unexpected that they resented. Having established our habits, birds in general insist that we shall govern ourselves by them, and not depart from our accustomed orbit.

On near acquaintance, I found the jay possessed of a vocabulary more copious than that of any other bird I know, though the flicker does not lack variety of expression. When some aspiring scientist is ready to study the language of birds, I advise him to experiment with the blue jay. He is exceedingly voluble, always ready to talk, and not in the least backward in exhibiting his accomplishments. The low-toned, plaintive sounding conversation of the jays with each other, not only beside the nest, but when flying together or apart, or in brief interviews in the lilac bush, pleased me especially, because it was exactly the same prattle that a pet blue jay was accustomed to address to me; and it confirmed what I had always believed from his manner, that it was his most loving and intimate expression, the tone in which he addresses his best beloved.

Beside the well-known squawk, which Thoreau aptly calls "the brazen trump of the impatient jay," the shouts and calls and war-cries of the bird can hardly be numbered, and I have no doubt each has its definite meaning. More rarely may be heard a clear and musical two-note cry, sounding like "ke-lo! ke-lo!" This seems to be something special in the jay language, for not only is it peculiar and quite unlike every other utterance, but I never saw the bird when he delivered it, and I was long in tracing it home to him. Aside from the cries of war and victory, jays have a great variety of notes of distress; they can put more anguish and despair into their tones than any other living creature of my acquaintance. Some, indeed, are so moving that the sympathetic hearer is sure that, at the very least, the mother's offspring are being murdered before her eyes; and on rushing out, prepared to risk his life in their defense, he finds, perhaps, that a child has strayed near the tree, or something equally dreadful has occurred. Blue Jays have no idea of relative values; they could not make more ado over a

heart-breaking calamity than they do over a slight annoyance. Some of their cries, notably that of the jay baby,
sound like the wail of a human infant. As to one curious
utterance in the jay répertoire, I could not quite make up
my mind whether it was a real call to arms, or intended
as a joke on the neighborhood. When a bird, without visible provocation, suddenly burst out with this loud two-
note call, instantly every feathered individual was on the
alert,—sprang to arms, as it were. Blue jays joined in, robins hurried to the tops of the tallest trees and added
their excited notes, with jerking wings and tail, and at
the second or third repetition the whole party precipi-
tated itself as one bird—upon what? Nothing that I could
discover.

XII. STORY OF THE NESTLINGS

While I was studying the manners and customs of the bird in blue, babies were growing up in the pine-tree nest. Five days after I began to observe, I saw little heads above the edge. On the sixth day they began, as mothers say, to "take notice," stirring about in a lively way, clambering up into sight, and fluttering their draperies over the edge. Now came busy and hungry times in the jay family; the mother added her forces, and both parents worked industriously from morning till night.

On the seventh day I was up early, as usual, and, also as usual, my first act was to admire the view from my window. I fancied it was the most beautiful in the early morning, when the sun, behind the rampart of locust and other trees, threw the yard into deep shade, painting a thousand shadow pictures on the grass; but at still noon, when every perfect tree stood on its own shadow, openings looked dark and mysterious, and a bird was lost in the depths, then I was sure it was never so lovely; again at night, when wrapped in darkness, and all silent except the subdued whisper of the pine, with its

"Sound of the Sea,
O mournful tree,
In thy boughs forever clinging,"

I knew it could not be surpassed. I was up early, as I said, when the dove was cooing to his mate in the distance, and before human noises had begun, and then I heard the baby cry from the pine-tree,—a whispered jay squawk, constantly repeated.

On this day the first nestling mounted the edge of his high nursery, and fluttered his wings when food approached. Every night after that it grew more and more difficult to settle the household in bed, for everybody wanted to be on top; and no sooner would one arrange himself to his mind than some "under one," not relishing his crushed position, would struggle out, step over his brothers and sisters, and take his place on top, and then the whole thing would have to be done over. I think that mamma had often to put a peremptory end to these difficulties by sitting down on them, for frequently it was a very turbulent-looking nest when she calmly placed herself upon it.

Often, in those days, I wished I could put myself on a level with that little castle in the air, and look into it, filled to the brim with beauty as I knew it was. But I had not long to wait, for speedily it became too full, and ran over into the outside world. On the eighth day one ambitious youngster stepped upon the branch beside the nest and shook himself out, and on the ninth came the plunge into the wide, wide world. While I was at breakfast he made his first effort, and on my return I saw him on a branch about a foot below the nest, the last step on papa's winding stair. Here he beat his wings and plumed himself vigorously, rejoicing, no doubt, in his freedom and in plenty of room. Again and again he nearly lost his balance, in his violent attempts to dress his beautiful plumage, and remove the last remnant of nest mussiness. But he did not fall, and at last he began to look about him. One cannot but wonder what he thought when he

"First opened wondering eyes and found
A world of green leaves all around,"

looking down upon us from his high perch, complete to the little black necklace, and lacking only length of tail of being as big as his parents.

After half an hour of restless putting to rights, the little jay sat down patiently to wait for whatever might come to him. The wind got up and shook him well, but he rocked safely on his airy seat. Then some one approached. He leaned over with mouth open, and across the yard I heard his coaxing voice. But alas! though he was on the very threshold, the food-bearer omitted that step, and passed him by. Then the little one looked up wistfully, apparently conscious of being at a disadvantage. Did he regret the nest privileges he had abandoned? Should he retrace his steps and be a nestling? That the thought passed through his head was indicated by his movements. He raised himself on his legs, turned his face to his old home, and started up, even stepped one small twig nearer. But perish the thought! he would not go back! He settled himself again on his seat.

All things come in time to him who can wait, and the next provision stopped at the little wanderer. His father alighted beside him and fed him two mouthfuls. Thus fortified, his ambition was roused, and his desire to see more, to do more. He began to jump about on his perch, facing first this way, then that; he crept to the outer end of the branch he was on, and was lost to view behind a thick clump of pine needles. In a few minutes he returned, considered other branches near, and, after some study, did really go to the nearest one. Then, step by step, very deliberately, he mounted the winding stair of his father, using, however, every little twig that the elder had vaulted over at a bound. Finally he reached the branch opposite his birthplace, only the tree-trunk between. The trunk was small, home was invitingly near, he was tired; the temptation was too great, and in a minute he was cuddled down with his brothers, having been on a journey of an hour. In the nest, all this time, there had been a hurry and skurry of dressing, as though the house were to be vacated, and no one wished to be late. After a rest and probably a nap, the ambitious young jay took a longer trip: he flew to the next tree, and, I believe, returned no more.

The next day was spent by all the nestlings in hopping about the three branches on which their home was built, making beautiful pictures of themselves every moment; but whenever the bringer of supplies drew near, each little one hastened to scramble back to the nest, to be ready for his share. The last day in the old home had now arrived. One by one the birdlings flew to the maple, and turned their backs on their native tree forever; and that night the "mournful tree" was entirely deserted.

The exit was not accomplished without its excitement. After tea, as I was congratulating myself that they were all safely out in the world, without accident, suddenly there arose a terrible outcry, robin and blue jay voices in chorus. I looked over to the scene of the fray, and saw a young jay on the ground, and the parents frantic with anxiety. Naturally, my first impulse was to go to their aid, and I started; but I was saluted with a volley of squawks that warned me not to interfere. I retired meekly, leaving the birds to deal with the difficulty as they best could, while from afar I watched the little fellow as he scrambled around in the grass. He tried to fly, but could not rise more than two feet. Both the elders were with him, but seemed unable to help him, and night was coming on. I resolved, finally, to "take my life in my hands," brave those unreasoning parents, and place the infant out of the way of cats and boys.

As I reached the doorstep I saw that the youngster had begun to climb the trunk of a locust-tree. I stood in amazement and saw that baby climb six feet straight up the trunk. He did it by flying a few inches, clinging to the bark and resting, then flying a few inches more. I watched, breathless, till he got nearly to the lowest branch, when alas! his strength or his courage gave out, and he fell back to the ground. But he pulled himself together, and after a few minutes more of struggling through the grass he came to the trunk of the maple next his native pine. Up this he went in the same way, till he reached a branch, where I saw him sitting with all the dignity of a young jay (old jays have no dignity). While he

was wrestling with fate and his life was in the balance,
the parents had kept near him and perfectly silent, unless some one came near, when they filled the air with
squawks, and appeared so savage that I honestly believe
they would have attacked any one who had tried to lend a
hand.

But still the little blue-coat had not learned sufficient modesty of endeavor, for the next morning he
found himself again in the grass. He tried climbing, but
unfortunately selected a tree with branches higher than
he could hold out to reach; so he fell back to the ground.
Then came the inexorable demands of breakfast, with
which no one who has been up since four o'clock will decline to comply. On my return, the straggler was mounted on a post that held a tennis net, three or four feet
from the ground. One of the old birds was on the rope
close by him, and there I left them. Once more I saw him
fall, but I concluded that since he had learned to climb,
and the parents would not accept my assistance any way,
he must take care of himself. I suppose he was the
youngest of the brood, who could not help imitating his
elders, but was not strong enough to do as they did. On
the following day he was able to keep his place, and he
came to the ground no more.

From that day I saw, and, what was more evident,
heard the jay babies constantly, though they wandered
far from the place of their birth. Their voices waxed
stronger day by day; from morning to night they called
vigorously; and very lovely they looked as they sat on the
branches in their brand-new fluffy suits, with their tails
a little spread, and showing the snowy borderings beau-
tifully. Twenty-two days after they bade farewell to the

old home before my window they were still crying for food, still following their hard-working parents, and, though flying with great ease, never coming to the ground (that I could see), and apparently having not the smallest notion of looking out for themselves.

XIII. BLUE JAY MANNERS

Early in my acquaintance with the jay family, wishing to induce the birds of the vicinity to show themselves, I procured a quantity of shelled corn, and scattered a few handfuls under my window every night. This gave me opportunity to note, among other things, the jay's way of conducting himself on the ground, and his table manners. To eat a kernel of dry corn, he flew with it to a small branch, placed it between his feet (the latter of course being close together), and, holding it thus, drew back his head and delivered a blow with that pickaxe beak of his that would have broken a toe if he had missed by the shadow of an inch the grain for which it was intended. I was always nervous when I saw him do it, for I expected an accident, but none ever happened that I know of. When the babies grew clamorous all over the place, the jay used to fill his beak with the whole kernels. Eight were his limit, and those kept the mouth open, with one sticking out at the tip. Thus loaded he flew off, but was back in two minutes for another supply. The red-headed woodpecker, who claimed to own the corn-field, seemed to think this a little grasping, and protested against such a wholesale performance; but the over-worked jay simply jumped to one side when he came at him, and went right on picking up corn. When he had time to spare from his arduous duties, he sometimes indulged his passion for burying things by carrying a grain off on the lawn with an air of most important business, and driving it into the ground, hammering it well down out of sight.

The blue jay's manner of getting over the ground was peculiar, and especially his way of leaving it. He proceeded by high hops, bounding up from each like a rubber ball; and when ready to fly he hopped farther and bounded higher each time, till it seemed as if he were too high to return, and so took to his wings. That is exactly the way it looked to an observer; for there is a lightness, an airiness of bearing about this apparently heavy bird impossible to describe, but familiar to those who have watched him.

Some time after the blue jay family had taken to roaming about the grounds, I had a pleasing little interview with one of them in the raspberry patch. This was a favorite resort of the neighboring birds, where I often betook myself to see who came to the feast. This morning I was sitting quietly under a spruce-tree, when three blue jays came flying toward me with noise and outcries, evidently in excitement over something. The one leading the party had in his beak a white object, like a piece of bread, and was uttering low, complaining cries as he flew; he passed on, and the second followed him; but the third seemed struck by my appearance, and probably felt it his duty to inquire into my business, for he alighted on a tree before me, not ten feet from where I sat. He began in the regular way, by greeting me with a squawk; for, like some of his bigger (and wiser?) fellow-creatures, he assumed that a stranger must be a suspicious personage, and an unusual position must mean mischief. I was very comfortable, and I thought I would see if I could not fool him into thinking me a scarecrow, companion to those adorning the "patch" at that moment. I sat motionless, not using my glass, but looking him squarely in the eyes. This seemed to impress him; he ceased squawking, and hopped a twig nearer, stopped, turned one calmly obser-vant eye on me, then quickly changed to the other, as if to see if the first had not deceived him. Still I did not move, and he was plainly puzzled to make me out. He came nearer and nearer, and I moved only my eyes to keep them on his. All this time he did not utter a sound, but studied me as closely, and to all appearances as care-fully, as ever I had studied him. Obviously he was in doubt what manner of creature it was, so like the human race, yet so unaccountably quiet. He tried to be uncon-cerned, while still not releasing me from strict surveil-lance; he dressed his feathers a little, uttering a soft whisper to himself, as if he said, "Well, I never!" then looked me over again more carefully than before. This pantomime went on for half an hour or more; and no one who had looked for that length of time into the eyes of a blue jay could doubt his intelligence, or that he had his thoughts and his well-defined opinions, that he had studied his observer very much as she had studied him, and that she had not fooled him in the least.

The little boy blue is one of the birds suffering under a bad name whom I have wished to know better, to see if perchance something might be done to clear up his reputation a bit. I am not able to say that he never steals the eggs of other birds, though during nearly a month of hard work, when, if ever, a few eggs would have been a welcome addition to his resources, and sparrows were sitting in scores on the place, I did not see or hear anything of the sort. I have heard of his destroying the nest, and presumably eating the eggs or young of the English sparrow, but the hundred or two who raised their broods and squawked from morning to night in the immediate vicinity of the pine-tree household never intimated that they were disturbed, and never showed hostility to their neighbors in blue. Moreover, there is undoubtedly something to be said on the jay's side. Even if he does indulge in these little eccentricities, what is he but a "collector"? And though he does not claim to be working "in the interest of science," which bigger collectors invariably do, he is working in the interest of life, and life is more than science. Even a blue jay's life is to him as precious as ours to us, and who shall say that it is not as useful as many of ours in the great plan?

The only indications of hostilities that I observed in four weeks' close study, at the most aggressive time of bird life, nesting-time, I shall relate exactly as I saw them, and the record will be found a very modest one. In this case, certainly, the jay was no more offensive than the meekest bird that has a nest to defend, and far less belligerent than robins and many others. On one occasion a strange blue jay flew up to the nest in the pine. I could not discover that he had any evil intention, except just to see what was going on, but one of the pair flew at him with loud cries, which I heard for some time after the two had disappeared in the distance, and when our bird returned, he perched on an evergreen, bowing and "jouncing" violently, his manner plainly defying the enemy to "try it again." At another time I observed a savage fight, or what looked like it, between two jays. I happened not to see the beginning, for I was particularly struck that morning with the behavior of a bouquet of nasturtiums which stood in a vase on my table. I never was fond of these flowers, and I noticed then for the first time how very self-willed and obstinate they were. No matter how nicely they were arranged, it would not be an hour before the whole bunch was in disorder, every blossom turning the way it preferred, and no two looking in the same direction. I thought, when I first observed this, that I must be mistaken, and I took them out and rearranged them as I considered best; but the result was always the same, and I began to feel that they knew altogether too much for their station in the vegetable world. I was trying to see if I could discover any method in their movements, when I was startled by a flashing vision of blue down under the locusts, and, on looking closely, saw two jays flying up like quarrelsome cocks,—only not together, but alternately, so that one was in the air all the time. They flew three feet high, at least, all their feathers on end, and looking more like shapeless masses of blue feathers than like birds. They did not pause or rest till one seemed to get the other down. I could not see from my window well enough to be positive, but both were in the grass together, and only one in sight, who stood perfectly quiet. He appeared to be holding the other down, for occasionally there would be a stir below, and renewed vigilance on the part of the one I could see. Several minutes passed. I became very uneasy. Was he killing him? I could stand it no longer, so I ran down. But my coming was a diversion, and both flew. When I reached the place, one had disappeared, and the other was hopping around the tree in great excitement, holding in his beak a fluffy white feather about the size of a jay's breast feather. I did not see the act, and I cannot absolutely declare it, but I have no doubt that he pulled that feather from the breast of his foe as he held him down; how many more with it I could not tell, for I did not think of looking until it was too late.

Again one day, somewhat later, when blue jay and catbird babies were rather numerous, I saw a blue jay dive into a lilac bush much frequented by catbirds, young and old together. Instantly there arose a great cry of distress, as though some one were hurt, and a rustling of leaves, proclaiming that a chase, if not a fight, was in progress. I hurried downstairs, and as I appeared the jay flew, with two catbirds after him, still crying in a way I had never heard before. I expected nothing less than to find a young catbird injured, but I found nothing. Whether the blue jay really had touched one, or it was a mere false alarm on the part of the very excitable catbirds, I could not tell. This is the only thing I have seen in the jay that might have been an interference with another bird's rights; and the catbirds made such a row when I came near their babies that I strongly suspect the only guilt of the jay was alighting in the lilac they had made their headquarters.

The little boy blue in the apple-tree, already spoken of, did not get his family off with so little adventure as his pine-tree neighbor. The youngling of this nest came to the ground and stayed there. The people of the house returned him to the tree several times, but every time he fell again. Three or four days he wandered about the neighborhood, the parents rousing the country with their uproar, and terrorizing the household cat to such a point of meekness that no sooner did a jay begin to squawk than he ran to the door and begged to come in. At last, out of mercy, the family took the little fellow into the house, when they saw that he was not quite right in some way. One side seemed to be nearly useless; one foot did not hold on; one wing was weak; and his breathing seemed to be one-sided. The family, seeing that he could not take care of himself, decided to adopt him. He took kindly to human care and human food, and before the end of a week had made himself very much at home. He knew his food provider, and the moment she entered the room he rose on his weak little legs, fluttered his wings violently, and presented a gaping mouth with the jay baby cry issuing therefrom. Nothing was ever more droll than this sight. He was an intelligent youngster, knew what he wanted, and when he had had enough. He would eat bread up to a certain point, but after that he demanded cake or a berry, and his favorite food was an egg. He was exceedingly curious about all his surroundings, examined everything with great care, and delighted to look out of the window. He selected his own sleeping-place,—the upper one of a set of bookshelves,—and refused to change; and he watched the movements of a wounded woodcock as he ran around the floor with as much interest as did the people. Under human care he grew rapidly stronger, learned to fly more readily and to use his weak side; and every day he was allowed to fly about in the trees for hours. Once or twice, when left out, he returned to the house for food and care; but at last came a day when he returned no more. No doubt he was taken in charge again by his parents, who, it was probable, had not left the neighborhood.

After July came in, and the baby blue jays could hardly be distinguished from their parents, my studies took me away from the place nearly all day, and I lost sight of the family whose acquaintance had made my June so delightful.

XIV. THE GREAT CAROLINIAN

All through June of that summer I studied the birds in the spacious inclosure around my "Inn of Rest." But as that month drew near its end,

> "The happy birds that change their sky
> To build and brood, that live their lives
> From land to land,"

almost disappeared. Blue jay babies wandered far off, where I could hear them it is true, but where—owing to the despair into which my appearance threw the whole jay family—I rarely saw them; orchard and Baltimore orioles had learned to fly, and carried their ceaseless cries far beyond my hearing; catbirds and cardinals, doves and golden-wings, all had raised their broods and betaken themselves wherever their fancy or food drew them, certainly without the bounds of my daily walks. It was evident that I must seek fresh fields, or remove my quarters to a more northerly region, where the sun is less ardent and the birds less in haste with their nesting.

Accordingly I sought a companion who should also be a guide, and turned my steps to the only promising place in the vicinity, a deep ravine, through which ran a little stream that was called a river, and dignified with a river's name, yet rippled and babbled, and conducted itself precisely like a brook.

The Glen, as it was called, was a unique possession for a common work-a-day village in the midst of a good farming country. Long ago would its stately trees have been destroyed, its streamlet set to turning wheels, and Nature forced to express herself on those many acres, in corn and potatoes, instead of her own graceful and varied selection of greenery; or, mayhap, its underbrush cut out, its slopes sodded, its springs buried in pipes and put to use, and the whole "improved" into dull insipidity,—all this, but for the will of one man who held the title to the grounds, and rated it so highly, that, though willing to sell, no one could come up to his terms. Happy delusion! that blessed the whole neighborhood with an enchanting bit of nature untouched by art. Long may he live to keep the deeds in his possession, and the grounds in their own wild beauty.

The place was surrounded by bristling barbed fences, and trespassers were pointedly warned off, so when one had paid for the privilege, and entered the grounds, he was supposed to be safe from intrusion, except of others who had also bought the right. The part easily accessible to hotel and railroad station was the scene of constant picnics, for which the State is famous, but that portion which lay near my place of study was usually left to the lonely kingfisher—and the cows. There the shy wood dwellers set up their households, and many familiar upland birds came with their fledglings; that was the land of promise for bird-lovers, and there one of them decided to study.

We began with the most virtuous resolves. We would come at five o'clock in the morning; we would catch the birds at their breakfast. We did; it was a lovely morning after a heavy rain, on which we set out to explore the ravine for birds. The storm in passing had taken the breeze with it, and not a twig had stirred since. Every leaf and grass blade was loaded with rain-drops. Walking in the grass was like wading in a stream; to touch a bush was to evoke a shower. But though our shoes were wet through, and our garments well sprinkled, before we reached the barbed fence, over or under or through or around which we must pass to our goal, we would not be discouraged; we went on.

As to the fence, let me, in passing, give my fellow drapery-bearers a hint. Carry a light shawl, or even a yard of muslin, to lay across the wire you can step over (thus covering the mischievous barbs), while a good friend holds up with strong hand the next wire, and you slip through. Thus you may pass this cruel device of man without accident.

Having circumvented the fence, the next task was to descend the steep sides of the ravine. The difficulty was, not to get down, for that could be done almost anywhere, but to go right side up; to land on the feet and not on the head was the test of sure-footedness and climbing ability. We conquered that obstacle, cautiously creeping down rocky steps, and over slippery soil, steadying ourselves by bushes, clasping small tree-trunks, scrambling over big ones that lay prone upon the ground, and thus we safely reached the level of the stream. Then we passed along more easily, stooping under low trees, crossing the beds of tiny brooks, encircling clumps of shrubbery (and catching the night's cobwebs on our faces), till we reached a fallen tree-trunk that seemed made for resting. There we seated ourselves, to breathe, and to see who lived in the place.

One of the residents proclaimed himself at once,

> "To left and right
> The cuckoo told his name to all the hills,"—

and in a moment we saw him, busy with his breakfast. His manner of hunting was interesting; he stood perfectly still on a branch, his beak pointed upward, but his head so turned that one eye looked downward. When something attracted him, he almost fell off his perch, seized the morsel as he passed, alighted on a lower branch, and at once began looking around again. There was no frivolity, no flitting about like a little bird; his conduct was grave and dignified, and he was absolutely silent, except when at rare intervals he mounted a branch and uttered his call, or song, if one might so call it. He managed his long tail with grace and expression, holding it a little spread as he moved about, thus showing the white tips and "corners."

While we were absorbed in cuckoo affairs the sun peeped over the trees, and the place was transfigured. Everything, as I said, was charged with water, and looking against the sun, some drops hanging from the tip of a leaf glowed red as rubies, others shone out blue as sapphires, while here and there one scintillated with many colors like a diamond, now flashing red, and now yellow or blue.

> "The humblest weed
> Wore its own coronal, and gayly bold
> Waved jeweled sceptre."

In that spot we sat an hour, and saw many birds, with whom it was evidently a favorite hunting-ground. But no one seemed to live there; every one appeared to be passing through; and realizing as we did, that it was late in the season, our search for nests in use was rather half-hearted anyway. As our breakfast-time drew near we decided to go home, having found nothing we cared to study. Just as we were taking leave of the spot I heard, nearly at my back, a gentle scolding cry, and glancing around, my eyes fell upon two small birds running down the trunk of a walnut sapling. A few inches above the ground one of the pair disappeared, and the other, still scolding, flew away. I hastened to the spot—and there I found my great Carolinian.

The nest was made in a natural cavity in the side of a stump six or eight inches in diameter and a foot high. It seemed to be of moss, completely roofed over, and stooping nearer its level I saw the bird, looking flattened as if she had been crushed, but returning my gaze, bravely resolved to live or die with her brood. I noted her color, and the peculiar irregular line over her eye, and then I left her, though I did not know who she was. Nothing would have been easier than to put my hand over her door and catch her, but nothing would have induced me to do so—if I never knew her name. Time enough for formal introductions later in our acquaintance, I thought, and if it happened that we never met again, what did I care how she was named in the books?

I did not at first even suspect her identity, for who would expect to find the great Carolina wren a personage of less than six inches! even though he were somewhat familiar with the vagaries of name-givers, who call one bird after the cat, whom he in no way resembles, and another after the bull, to whom the likeness is, if possible, still less. What was certain was that the nest belonged to wrens, and was admirably placed for study; and what I instantly resolved was to improve my acquaintance with the owners thereof.

The little opening in the woods, which became the Wren's Court, when their rank was discovered, was a most attractive place, shaded enough to be pleasant, while yet leaving a goodly stretch of blue sky in sight, bounded on one side by immense forest trees—walnut, butternut, oak, and others—which looked as if they had stood there for generations; on the other side, the babbling stream, up and down which the kingfisher flew and clattered all day. One way out led to the thicket where a wood-thrush was sitting in a low tree, and the other, by the Path Difficult, up to the world above. The seat, across the court from the nest, had plainly been arranged by some kind fate on purpose for us. It was the trunk of a tree, which in falling failed to quite reach the ground, and so had bleached and dried, and it was shaded and screened from observation by vigorous saplings which had sprung up about it. The whole was indeed an ideal nook, well worthy to be named after its distinguished residents.

Thoreau was right in his assertion that one may see all the birds of a neighborhood by simply waiting patiently in one place, and into that charming spot came "sooner or later" every bird I had seen in my wanderings up and down the ravine. There sang the scarlet tanager every morning through July, gleaming among the leaves of the tallest trees, his olive-clad spouse nowhere to be seen, presumably occupied with domestic affairs. There the Acadian flycatcher pursued his calling, fluttering his wings and uttering a sweet little murmur when he alighted. Into that retired corner came the cries of flicker and blue jay from the high ground beyond. On the edge sang the indigo-bird and the Western wood-pewee, and cardinal and wood-thrush song formed the chorus to all the varied notes that we heard.

Upon our entrance the next morning, my first glance at the nest was one of dismay—the material seemed to be pulled out a little. Had it been robbed! had some vagabond squirrel thrust lawless paws into the little home! I looked closely; no, there sat, or rather there lay the little mother. But she did not relish this second call. She flew, fluttering and trailing on the ground, as if hurt, hoping, of course, to attract us away from her nest. Seeing that of no avail, however, which she quickly did, she retreated to a low branch, threw back her head, and uttered a soft "chur-r-r," again and again repeated, doubtless to her mate. But that personage did not make his appearance, and we examined the nest. There were five eggs, white, very thickly and evenly specked with fine dots of dark color. An end of one that stuck up was plain white, perhaps the others were the same; we did not inquire too closely, for what did we care for eggs, except as the cradles of the future birds?

Very soon we retired to our seat across the court and became quiet, to wait for what might come. Suddenly, with almost startling effect,

> "A bird broke forth and sung
> And trilled and quavered and shook his throat."

It was a new voice to us, loud and clear, and the song, consisting of three clauses, sounded like "Whit-e-ar! Whit-e-ar! Whit-e-ar!" then a pause, and the same repeated, and so on indefinitely. It came nearer and still nearer, and in a moment we saw the bird, a tiny creature, red-brown on the back, light below—the image of the little sitter in the stump, as we remarked with delight; we hoped he was her mate. He did not seem inclined to go to the nest, but stayed on a twig of a dead branch which hung from a large tree near by.

While the stranger was pouring out his rhapsody, head thrown back, tail hanging straight down, and wings slightly drooped, I noticed a movement by the nest, and fixed my eyes upon that. The little dame had stolen out of her place, and now began the ascent of the sapling which started out one side of her small stump. Up the trunk she went with perfect ease, running a few steps, and then pausing a moment before she took the next half-dozen. She did not go bobbing up like a woodpecker, nor did she steady herself with her tail, like that frequenter of tree-trunks; she simply ran up that almost perpendicular stick as a fly runs up the wall. Meanwhile her mate, if that he were, kept up his ringing song, till she reached the top of the sapling, perhaps seven or eight feet high, and flew over near him. In an instant the song ceased, and the next moment two small birds flew over our heads, and we heard chatting and churring, and then silence.

Without this hint from the wren we should rarely have seen her leave the nest; we should naturally have watched for wings, and none might come or go, while she was using her feet instead. She returned in the same way; flying to the top, or part way up her sapling, she ran down to her nest as glibly as she had run up. The walnut-trunk was the ladder which led to the outside world. This pretty little scene was many times repeated, in the days that we spent before the castle of our Carolinians; the male announcing himself afar with songs, and approaching gradually, while his mate listened to the notes that had wooed her, and now again coaxed her away from her sitting, for a short outing with him. Sometimes, though rarely, she came out without this inducement, but during her sitting days she usually went only upon his invitation.

Before many days we had fully identified the pair. The song had puzzled me at first, for though extraordinary in volume for a bird of his size, and possessing that indefinable wren quality, that abandon and unexpectedness, as if it were that instant inspired, it had yet few notes, and I missed the exquisite tremolo that makes the song of the winter-wren so bewitching. But I "studied him up," and learned that his finest and most characteristic song is uttered in the spring only. After nesting has begun, he gives merely these musical calls, which, though delightful, do not compare—say the books—with his ante-nuptial performance. I was too late for that, but I was glad and thankful for these.

Moreover, the wren varied his songs as the days went on. There were from two to five notes in a clause, never more, and commonly but three. This clause he repeated again and again during the whole of one visit; but the next time he came he had a new one, which likewise he kept to while he stayed. Again, when, some days later, he took part in feeding, he frequently changed the song as he left the nest. Struck by the variety he gave to his few notes, after some days I began to take them down in syllables as they expressed themselves to my ear, for they were sharp and distinct. Of course, these syllables resemble his sound about as a dried flower resembles the living blossom, but they serve the same purpose, to reproduce them in memory. In that way I recorded in three days eighteen different arrangements of his notes. Doubtless there were many more; indeed, he seemed to delight in inventing new combinations, and his taste evidently agreed with mine, for when he succeeded in evolving a particularly charming one, he did not easily change it. One that specially pleased me I put down as "Shame-ber-ee!" and this was his favorite, too, for after the day he began it, he sang it oftener than any other. It had a peculiarly joyous ring, the second note being a third below the first, and the third fully an octave higher than the second. I believe he had just then struck upon it, his enjoyment of it was so plain to see.

The Wren's Court was a distracting spot to study one pair of small birds. So many others came about, and always, it seemed, in some crisis in wren affairs, when I dared not take my eyes from my glass, lest I lose the sequence of events. There appeared sometimes to be a thousand whispering, squealing, and smacking titmice in the trees over my head, and a whole regiment of great-crested flycatchers and others on one side. I was glad I was familiar with all the flicker noises, or I should have been driven wild at these moments, so many, so various, and so peculiar were their utterances; likewise thankful that I knew the row made by the jay on the bank above was not a sign of dire distress, but simply the tragic manner of the family.

Again, when the wind blew, it was impossible to see the little folk that chattered and whispered and "dee-dee'd" overhead, and though we were absolutely certain a party of tufted tits and chickadees and black and white creepers, who always seemed to travel in company, were frolicking about, we could not distinguish them from the dancing and fluttering leaves.

When the day was favorable, and the wren had gone his way, foraging in silence over the low ground at our back, and an old stump that stood there, and the sitter had settled herself in her nest for another half hour, we could look about at whoever happened to be there. Thus I made further acquaintance with the great-crested flycatcher. Hitherto I had known these birds only as they travel through a neighborhood not their own, appearing on the tops of trees, and crying out in martial tones for the inhabitants to bring on their fighters, a challenge to all whom it may concern. It was a revelation, then, to see them quietly at home like other birds, setting up claims to a tree, driving strangers away from it, and spending their time about its foot, seeking food near the ground, and indulging in frolics or fights, whichever they might be, with squealing cries and a rushing flight around their tree. In the latter part of our study, the great-crest babies were out, noisy little fellows, who insisted on being fed as peremptorily as their elders demand their rights and privileges.

To make the place still more maddening for study, the birds seemed to sweep through the woods in waves. For a long time not a peep would be heard, not a feather would stir; then all at once

"The air would throb with wings,"

and birds would pour in from all sides, half a dozen at a time, making us want to look six ways at once, and rendering it impossible to confine ourselves to one. Then, after half an hour of this superabundance, one by one would slip out, and by the time we began to realize it, we were alone again.

We had watched the wren for nine days when there came an interruption. It happened thus: A little farther up the glen we had another study, a wood-thrush nest in a low tree, and every day, either coming or going, we were accustomed to spend an hour watching that. Our place of observation was a hidden nook in a pile of rocks, where we were entirely concealed by thick trees, through which, by a judicious thinning out of twigs and leaves, we had made peepholes, for the thrush mamma would not tolerate us in her sight. To reach our seats and not alarm the suspicious little dame, we always entered from the back, slowly and cautiously climbed the rocks by a rude path which already existed, and slipped in under cover of our leafy screen.

On the morning of the tenth day we entered the ravine from the upper end, and made our first call upon the thrush. We had been seated in silence for ten or fifteen minutes, and I was beginning to get uneasy because no bird came to the nest, when a diversion occurred that drove thrush affairs out of our minds. We heard footsteps! It must be remembered that we were alone in this solitary place, far from a house, and naturally we listened eagerly. The steps drew nearer, and then we heard loud breathing. We exchanged glances of relief—it was a cow! But while we were congratulating ourselves began a crashing of branches, a fiercer breathing, a rush, and a low bellow!

This was no meek cow! we turned pale,—at any rate we felt pale,—but we tried to encourage each other by suggesting in hurried whispers that he surely would not see us. Alas! the next instant he broke through the bushes, and to our horror started at once up our path to the rocks; in a moment he would be upon us! We rose hastily, prepared to sell our lives dearly, when, as suddenly as he had come, he turned and rushed back. Whether the sight of us was too much for his philosophy, or whether he had gone for reinforcements, we did not inquire. We instantly lost our interest in birds and birds' nests; we gathered up our belongings and fled, not stopping to breathe till we had put the barbiest of barbed wire fences between us and the foe.

Once outside, however, we paused to consider: To give up our study was not to be thought of; to go every day in fear and dread was equally intolerable. I wrote to the authorities of whom I had purchased the right to enter the place. They promptly denied the existence of any such animal on the premises. I replied to the effect that "seeing is believing," but they reaffirmed their former statement, assuring me that there were none but harmless cows in the glen. I did not want to waste time in an unprofitable correspondence, and I did want to see the wrens, and at last a bright thought came,—I would hire an escort, a country boy used to cattle, and warranted not afraid of them. I inquired into the question of day's wages, I looked about among the college students who were working their way to an education, and I found an ideal protector,—an intelligent and very agreeable young man, brought up on a farm, and just graduated, who was studying up mathematics preparatory to school-teaching in the fall. The bargain was soon made, and the next morning we started again for the glen, our guardian armed with his geometry and a big club. Three days, however, had been occupied in perfecting this arrangement, and I approached the spot with anxiety; indeed, I am always concerned till I see the whole family I am watching, after only a night's interval, and know they have survived the many perils which constantly threaten bird-life, both night and day.

XV. THE WRENLINGS APPEAR

The moment we entered the court I saw there was news. My eyes being attracted by a little commotion on a dogwood-tree, I saw a saucy tufted titmouse chasing with cries one of the wrens who had food in its beak. With most birds this proclaims the arrival of the young family as plainly as if a banner had been hung on the castle walls. Whether the tit was after the food, or trying to drive the wren off his own ground, we could not tell, nor did we much care; the important fact was that babies were out in the walnut-tree cottage. The food bearer went to the nest, and in a moment came up the ladder, so joyous and full of song that he could not wait to get off his own tree, but burst into a triumphant ringing "White-ar!" that must have told his news to all the world—who had ears to hear.

The mother did not at once give up her brooding, nor did I wonder when I peeped into the nest while she was off with her spouse, and saw what appeared to be five big mouths with a small bag of skin attached to each. Nothing else could be seen. She sat an hour at a time, and then her mate would come and call her off for a rest and a change, while he skipped down the ladder and fed the bairns. His way in this matter, as in everything else, was characteristic. He never went to the nest till he had called her off by his song. It was not till several days later, when she had given up brooding, that I ever saw the pair meet at the nest, and then it seemed to be accidental, and one of them always left immediately.

During the first few days the young parents came and went as of old, by way of the ladder, and I learned to know them apart by their way of mounting that airy flight of steps. He was more pert in manner, held his head and tail more jauntily, though he rarely pointed his tail to the sky, as do some of the wren family. He went lightly up in a dancing style which she entirely lacked, sometimes jumping to a small shoot that grew up quite near the walnut, and running up that as easily as he did the tree. Her ascent was of a business character; she was on duty, head and tail level with her body, no airs whatever. He was so full of happiness in these early days that frequently he could not take time to go to the top, but, having reached a height of two or three feet, he flew, and at once burst into rapturous song, even sang while flying over to the next tree. From this time they almost abandoned the ladder they had been so fond of, and flew directly to the nest from the ground, where they got all their food. This change was not because they were hard worked; I never saw birds who took family cares more easily. At the expiration of three days the mother brooded no more, and indeed it would have troubled her to find a place for herself, the nest was so full.

Every morning on entering the court I called at the nest, and always found five yellow beaks turned to the front. On the third day the heads were covered with slate-colored down; on the fourth, wing-feathers began to show among the heads, but the body was still perfectly bare; on the fifth, the eyes opened on the green world about them,—they were then certainly five days old, and may have been seven; owing to our unfortunate absence at the critical time I cannot be sure. On the seventh day the red-brown of the back began to show, and the white of the breast made itself visible, while the heads began to look feathery instead of fuzzy. Even then, however, they took no notice when I put my finger on them.

Long before this time the manner of the parents had changed. In the first place, they were more busy; foraging industriously on the ground, coming within ten or fifteen feet of us, without appearing to see us at all. In fact they had, after the first day, paid no attention to us, for we never had disturbed them, never went to the nest till sure that both were away, and kept still and quiet in our somewhat distant seat.

About this time they began to show more anxiety in their manner. The first exhibition was on the fourth day since we knew the young were hatched (and let me say that I believe they were just out of the shell the morning that we found the father feeding). On this fourth day the singer perched near the nest-tree, three or four feet from the ground, and began a very loud wren "dear-r-r-r! dear-r-r-r! dear-r-r-r!" constantly repeated. He jerked himself about with great apparent excitement, looking always on the ground as if he saw an enemy there. We thought it might be a cat we had seen prowling about, but on examination no cat was there. Gradually his tone grew lower and lower, and he calmed down so far as a wren can calm, though he did not cease his cries. I did not know he could be still so long, but I learned more about wren possibilities in that line somewhat later.

During this performance his mate came with food in her beak, and evidently saw nothing alarming, for she went to the nest with it. Still he stood gazing on the ground. Sometimes he flew down and returned at once, then began moving off, a little at a time, still crying, exactly as though he were following some one who went slowly. The call, when low, was very sweet and tender; very mournful too, and we got much wrought up over it, wishing—as bird students so often do—that we could do something to help. He was roused at last by the intrusion of a bird into his domain, and his discomfiture of this foe seemed to dispel his unhappy state of mind, for he at once broke out in joyous song, to our great relief. That was not the last exhibition of the wren's idiosyncrasy; he repeated it day after day, and finally he went so far as to interpolate low "dear-r-r's" into his sweetest songs. Perhaps that was his conception of his duty as protector to the family; if so, he was certainly faithful in doing it. It was ludicrously like the attitude of some people under similar circumstances.

While the young father was manifesting his anxiety in this way, the mother showed hers in another; she took to watching, hardly leaving the place at all. When she had her babies well fed for the moment, she went up the trunk a little, in a loitering way that I had never seen her indulge in before,—and a loitering wren is a curiosity. It was plain that she simply wished to pass away the time. She stepped from the trunk upon a twig on one side, stayed a little while, then passed to one on the other side, lingered a few moments, and so she went on. When she arrived at the height of two feet she perched on a small dead twig, and remained a long time—certainly twenty minutes—absolutely motionless. It was hard to see her, and if I had not watched her progress from the first, I should not have suspected her presence. A leaf would hide her, even the crossing of two twigs was ample screen, and when she was still it was hopeless to look for her. The only way we were able to keep track of either of the pair was by their incessant motions.

The Great Carolinian had a peculiar custom which showed that his coming with song was a ceremony he would not dispense with. He would often start off singing, gradually withdraw till fifty or seventy-five feet away, singing at every pause, and then, if one watched him closely, he might see him stop, drop to the ground, and hunt about in silence. When he was ready to come again, he would fly quietly a little way off, and then begin his singing and approaching, as if he had been a mile away. He never sang when on the ground after food, but so soon as he finished eating, he flew to a perch at least two feet high, generally between six and ten, and sometimes as high as twenty feet, and sang.

After a day or two of the wren's singular uneasiness, we discovered at least one object of his concern. It was a chipmunk, whom we had often noticed perched on the highest point of the little ledge of rocks near the nest. He seemed to be attending strictly to his own affairs, but after a good deal of "dear-r-r"-ing, the wren flew furiously at him, almost, if not quite, hitting him, and doing it again and again. The little beast did not relish this treatment and ran off, the bird following and repeating the assault. This was undoubtedly the foe that he had been troubled about all the time.

On the tenth or eleventh day of their lives (as I believe) I examined the babies in the nest a little more closely than before. I even touched them with my finger on head and beak. They looked sleepily at me, but did not resent it. If the mother were somewhat bigger, I should suspect her of giving them "soothing syrup," for they had exactly the appearance of being drugged. They were not overfed; I never saw youngsters so much let alone. The parents had nothing like the work of the robin, oriole, or blue jay. They came two or three times, and then left for half an hour or more, yet the younglings were never impatient for food.

The morning that the young wrens had reached the age of twelve days (that we knew of) was the 22d of July, and the weather was intensely warm. On the 21st we had watched all day to see them go, sure that they were perfectly well able. Obviously it is the policy of this family to prepare for a life of extraordinary activity by an infancy of unusual stillness. Never were youngsters so perfectly indifferent to all the world. In storm or sunshine, in daylight or darkness, they lay there motionless, caring only for food, and even that showed itself only by the fact that all mouths were toward the front. The under one of the pile seemed entirely contented to be at the bottom, and the top ones not to exult in their position; in fact, so far as any show of interest in life was concerned, they might have been a nestful of wooden babies.

On this morning, as we dragged ourselves wearily over the hot road to the ravine, we resolved that no handful of wrenlings should force us over that road again. Go off this day they should, if—as my comrade remarked —"we had to raise them by hand." My first call was at the nest, indifferent whether parents were there or not, for I had become desperate. There they lay, lazily blinking at me, and filling the nest overfull. The singer came rushing down a branch, bristled up, blustering, and calling "Dear-r-r-r!" at me, and I hoped he would be induced to hurry up his very leisurely brood.

We took our usual seats and waited. Both parents remained near the homestead, and little singing was indulged in; this morning there was serious business on hand, as any one could see. We were desirous of seeing the first sign of movement, so we resolved to cut away the last few leaves that hid the entrance to the nest. We had not done it before, partly not to annoy the birds, and partly not to have them too easily discovered by prowlers.

Miss R—— went to the stump, and cut away half a dozen leaves and twigs directly before their door. The young ones looked at her, but did not move. Then, as I had asked her to do, she pointed a parasol directly at the spot, so that I, in my distant seat, might locate the nest exactly. This seemed to be the last straw that the birdlings could endure; two of them flew off. One went five or six feet away, the other to the ground close by. Then she came away, and we waited again. In a moment two more ventured out and alighted on twigs near the nest. Then the mother came home, and acted as surprised as though she had never expected to have them depart. She went from a twig beside the tree to the nest, and back, about a dozen times, as if she really could not believe her eyes.

Anxious to see everything that went on, we moved our seats nearer, but this so disconcerted the pair that we did not stay long. It was long enough to hear the wren baby-cry, a low insect-like noise, and to see something that surprised and no less disgusted me, namely, every one of those babies hurry back to the tree, climb the trunk, and scramble back into the nest!—the whole exit to be begun again! It could not be their dislike of the "cold, cold world," for a cold world would be a luxury that morning.

Of any one who would go back into that crowded nest, with the thermometer on the rampage as it was then, I had my opinion, and I began to think I didn't care much about wrens anyway; we stayed, however, as a matter of habit, and I suppose they all had a nap after their tremendous exertion. But they manifestly got an idea into their heads at last, a taste of life. After a proper amount of consideration, one of the nestlings took courage to move again, and went so far as a twig that grew beside the door, looked around on the world from that post for a while, then hopped to another, and so on till he encircled the home stump. But when he came again in sight of that delectable nest, he could not resist it, and again he added himself to the pile of birds within. This youth was apparently as well feathered as his parents, and, except in length of tail, looked exactly like them; many a bird baby starts bravely out in life not half so well prepared for it as this little wren.

After nearly three hours of waiting, we made up our minds that these young folk must be out some time during the day, unless they had decided to take up permanent quarters in that hole in the stump, and what was more to the point, that the weather was too warm to await their very deliberate movements. So we left them, to get off the best way they could without us, or to stay there all their lives, if they so desired.

The nest, which at first was exceedingly picturesque—and I had resolved to bring it away, with the stump that held it—was now so demolished that I no longer coveted it. The last and sweetest song of the wren, "Shame-ber-ee!" rang out joyously as we turned our faces to the north, and bade a long farewell to the Great Carolinians.

XVI. THE APPLE-TREE NEST

All day long in the elm,
on their swaying perches swinging,
New-fledged orioles utter their restless, querulous notes.

Harriet Prescott Spofford

The little folk let out the secret, as little folk often do, and after they had called attention to it, I was surprised that I had not myself seen the pretty hammock swinging high up in the apple boughs.

It was, however, in a part of the grounds I did not often visit, partly because the trees close by, which formed a belt across the back of the place, grew so near together that not a breath of air could penetrate, and it was intolerable in the hot June days, and partly because my appearance there always created a panic. So seldom did a human being visit that neglected spot, that the birds did not look for guests, and a general stampede followed the approach of one.

On the eventful day of my happy discovery I was returning from my daily call upon a blue jay who had set up her home in an apple-tree in a neighbor's yard. The moment I entered the grounds I noticed a great outcry. It was loud; it was incessant; and it was of many voices. Following the sound, I started across the unmown field,

> "Through the bending grasses,
> Tall and lushy green,
> All alive with tiny things,
> Stirring feet and whirring wings
> Just an instant seen,"

and soon came in sight of the nest near the topmost twig of an old apple-tree.

It was about noon of a bright, sunny day, and I could see only that the nest was straw-color, apparently run over with little ones, and both the parents were industriously feeding. The cries suggested the persistence of young orioles, but it was not a Baltimore's swinging cradle, and the old birds were so shy, coming from behind the leaves, every one of which turned itself into a reflector for the sunlight, that I could not identify them.

Later in the day I paid them another visit, and finding a better post of observation under the shade of a sweet-briar bush, I saw at once they were orchard orioles, and that the young ones were climbing to the edge of the nest; I had nearly been too late!

Four o'clock was the unearthly hour at which I rose next morning to pursue my acquaintance with the little family in the apple-tree, fearful lest they should get the start of me. The youngsters were calling vociferously, and both parents were very busy attending to their wants and trying to stop their mouths, when I planted my seat before their castle in the air, and proceeded to inquire into their manners and customs. My call was, as usual, not received with favor. The mother, after administering the mouthful she had brought, alighted on a twig beside the nest and gave me a "piece of her mind." I admitted my bad manners, but I could not tear myself away. The anxious papa, very gorgeous in his chestnut and black suit, scenting danger to the little brood in the presence of the bird-student with her glass, at once abandoned the business of feeding, and devoted himself to the protection of his family,—which indeed was his plain duty. His way of doing this was to take his position on the tallest tree in the vicinity, and fill the serene morning air with his cry of distress, a two-note utterance, with a pathetic inflection which could not fail to arouse the sympathy of all who heard it. It was not excited or angry, but it proclaimed that here was distress and danger, and it had the effect of making me ashamed of annoying him. But I hardened my heart, as I often have to do in my study, and kept my seat. Occasionally he returned to the lower part of his own tree, to see if the monster had been scared or shamed away, but finding me stationary, he returned to his post and resumed his mournful cry.

At length the happy thought came to me that I might select a position a little less conspicuous, yet still within sight, so I moved my seat farther off, away back under a low-branched apple-tree, where a redbird came around with sharp "tsip's" to ascertain my business, and a catbird behind the briar-bush entertained me with delicious song. The oriole accepted my retirement as a compromise, and returned to his domestic duties, coming, as was natural and easiest, on my side of the tree. His habit was to cling to the side of the nest, showing his black and red-gold against it, while his mate alighted on the edge, and was seen a little above it. After feeding, both perched on neighboring twigs and looked about for a moment before the next food-hunting trip. I thought the father of the family exhibited an air of resignation, as if he concluded that, since the babies made so much noise, there was no use in trying longer to preserve the secret.

As a matter of fact, both our orioles need a good stock of patience as well as of resignation, for the infants of both are unceasing in their cries, and fertile in inventing variations in manner and inflection, that would deceive those most familiar with them. Two or three times in the weeks that followed, I rushed out of the house to find some very distressed bird, who, I was sure, from the cries, must be impaled alive on a butcher-bird's meathook, or undergoing torture at the hands—or beak of somebody. It was rather dangerous going out at that time (just at dusk), for it was the chosen hour for young men and maidens, of whom there were several, to wander about under the trees. Often, before I gave up going out at that hour, my glass, turned to follow a flitting wing, would bring before my startled gaze a pair of sentimental young persons, who doubtless thought I was spying upon them. My only safety was in directing my glass into the trees, where nothing but wings could be sentimental, and if a bird flitted below the level of branches, to consider him lost. On following up the cry, I always found a young oriole and a hard-worked father feeding him. The voice did not even suggest an oriole to me, until I had been deceived two or three times and understood it.

The young ones of the orchard oriole's nest lived up to the traditions of the family by being inveterate cry-babies, and making so much noise they could be heard far around. Sometimes their mother addressed them in a similar tone to their own, but the father resigned himself to the inevitable, and fed with dogged perseverance.

The apple-tree nest looked in the morning sun of a bright flax color, and two of the young were mounted on the edge, dressing their yellow satin breasts, and gleaming in the sunshine like gold.

A Baltimore oriole, passing over, seemed to be attracted by a familiar quality of sound, for he came down, alighted about a foot from the nest, and looked with interest upon the charming family scene. The protector of the pretty brood was near, but he kept his seat, and made no objections to the friendly call. Indeed, he flew away while the guest was still there, and having satisfied his curiosity, the Baltimore also departed upon his own business.

When the sun appeared over the tree-tops, he came armed with all his terrors. The breeze dwindled and died; the very leaves hung lifeless on the trees, and though, knowing that

> "Somewhere the wind is blowing,
> Though here where I gasp and sigh
> Not a breath of air is stirring,
> Not a cloud in the burning sky,"

the memory might comfort me, it did not in the slightest degree make me comfortable—I wilted, and retired before it. How the birds could endure it and carry on their work, I could not understand.

At noon I ventured out over the burning grass. The first youngster had left the nest, and was shouting from a tree perhaps twenty feet beyond the native apple. The others were fluttering on the edge, crying as usual. As is the customary domestic arrangement with many birds, the moment the first one flew, the father stopped coming to the nest, and devoted himself to the straggler, which was a little hard on the mother that hot day, for she had four to feed.

While I looked on, the second infant mustered up courage to start on the journey of life. A tall twig led from the nest straight up into the air, and this was the ladder he mounted. Step by step he climbed one leaf-stem after another, with several pauses to cry and to eat, and at last reached the topmost point, where he turned his face to the west, and took his first survey of the kingdoms of the earth. A brother nestling was close behind him, and the pretty pair, seeing no more steps above them, rested a while from their labors. In the mean time the first young oriole had gone farther into the trees, and papa with him.

The little dame worked without ceasing, though it must have been an anxious time, with nestlings all stirring abroad. I noticed that she fed oftenest the birdlings who were out, whether to strengthen them for further effort, or to offer an inducement to those in the nest to come up higher where food was to be had, she did not tell. I observed, also, that when she came home she did not, as before, alight on the level of the little ones, but above them. Perhaps this was to coax them upward; at any rate, it had that effect: they stretched up and mounted the next stem above, and so they kept on ascending. About three o'clock I was again obliged to surrender to the power of the sun, and retire for a season to a place he could not enter, the house.

Some hours passed before I made my next call, and I found that oriole matters had not rested, if I had; the two nestlings had taken flight to the tree the first one had chosen, and three were on the top twig above the nest, which latter swung empty and deserted. Mamma was feeding the three in her own tree, while papa attended as usual to the outsiders, and found leisure to drop in a song now and then.

While I watched, number three took his life in his hands (as it were) and launched out upon the air. He reached a tree not so far away as his brothers had chosen, and his mother sought him out and fed him there. But he did not seem to be satisfied with his achievement, or possibly he found the position rather lonely; at any rate, the next use of his wings was to return to his native apple, to the lower part. During this visit, the mother of the little brood, seeing, I suppose, her labors growing lighter, indulged herself and delighted me with a scrap of song, very sweet, as the song of the female oriole always is.

It was with forebodings that I approached the tree the next morning, foreboding speedily confirmed— the whole family was gone! Either I had not stayed late enough or I had not got up early enough to see the flitting; that song, meant something—it was my good-by.

Indeed it turned out to be my farewell, as I thought, for the whole tribe seemed to have vanished.

Usually it is not difficult to hunt up a little bird family in its wanderings, during the month following its leaving the nest, but this one I could neither see nor hear, and I was very sure those oriole babies had not so soon outgrown their crying; they must have been struck dumb or left the place.

Nearly three weeks later I was wandering about in what was called the glen, half a mile or more from where the apple-tree babies had first seen the light. It was a wild spot, a ravine, through which ran a stream, where many wood-birds sang and nested. On approaching a linden-tree loaded with blossoms, and humming with swarms of bees, I was saluted with a burst of loud song, interspersed with scolding. No one but an orchard oriole could so mix things, and sure enough! there he was, scrambling over the flowers. Something he found to his taste, whether the blossoms or the insects, I could not decide. On waiting a little, I heard the young oriole cry, much subdued since nesting days, and the tender "ye-ep" of the parent. The whole family was evidently there together, and I was very glad to see them once more.

The nest, which I had brought down, was a beautiful structure, made, I think, of very fine excelsior of a bright straw-color. It was suspended in an upright fork of four twigs, and lashed securely to three of them, while a few lines were passed around the fourth. Though it was in a fork, it did not rest on it, but was suspended three inches above it, a genuine hanging nest. It was three inches deep and wide, but drawn in about the top to a width of not more than two inches, with a bit of cotton and two small feathers for bedding. How five babies could grow up in that little cup is a problem. The material was woven closely together, and in addition stitched through and through, up and down, to make a firm structure. Around and against it hung still six apples, defrauded of their manifest destiny, and remaining the size of hickory-nuts. Three twigs that ran up were cut off, but the fourth was left, the tallest, the one sustaining the burden of the nest, and upon which the young birds, one after another, had mounted to take their first flight.

This pretty hammock, in its setting of leaves and apples, still swinging from the apple boughs, I brought home as a souvenir of a charming bird study.

XVII. CEDAR-TREE LITTLE FOLK

'T is there that the wild dove has her nest,
　And whenever the branches stir,
She presses closer the eggs to her breast,
　And her mate looks down on her.

Clare Beatrice Coffey

One of the voices that helped to make my June musical, and one more constantly heard than any other, was that of the

"Mourning dove who grieves and grieves,
　And lost! lost! lost! still seems to say,"

as the poet has it.

Now, while I dearly love the poets, and always long to enrich my plain prose with gems from their verse, it is sometimes a little embarrassing, because one is obliged to disagree with them. If they would only look a little into the ways of birds, and not assert, in language so musical that one can hardly resist it, that

"The birds come back to last year's nests,"

when rarely was a self-respecting bird known to shirk the labor of building anew for every family; or sing, with Sill,

"He has lost his last year's love, I know,"

when he did not know any such thing; and add,

"A thrush forgets in a year,"

which I call a libel on one of our most intelligent birds; or cry, with another singer,

"O voiceless swallow,"

when not one of the whole tribe is defrauded of a voice, and at least one is an exquisite singer; or accuse the nightingale of the superfluous idiocy of holding his (though they always say her) breast to a thorn as he sings, as if he were so foolish as to imitate some forms of human self-torture,—if they would only be a little more sure of their facts, what a comfort it would be to those who love both poets and birds!

No bird in our country is more persistently misrepresented by our sweet singers than the Carolina or wood dove—mourning dove, as he is popularly called; and in this case they are not to be blamed, for prose writers, even natural history writers, are quite as bad.

"His song consists," says one, "of four notes: the first seems to be uttered with an inspiration of the breath, as if the afflicted creature were just recovering its voice from the last convulsive sob of distress, and followed by three long, deep, and mournful moanings, that no person of sensibility can listen to without sympathy." "The solemn voice of sorrow," another writer calls it. All this is mere sentimentality, pure imagination; and if the writers could sit, as I have, under the tree when the bird was singing, they would change their opinion, though they would thereby lose a pretty and attractive sentiment for their verse. I believe there is

> "No beast or bird in earth or sky,
> Whose voice doth not with gladness thrill,"

though it may not so express itself to our senses. Certainly the coo of the dove is anything but sad when heard very near. It has a rich, far-off sound, expressing deep serenity, and a happiness beyond words.

First in the morning, and last at night, all through June, came to me the song of the dove. As early as four o'clock his notes began, and then, if I got up to look out on the lawn, where I had spread breakfast for him and other feathered friends, I would see him walking about with dainty steps on his pretty red toes, looking the pink of propriety in his Quaker garb, his satin vest smooth as if it had been ironed down, and quite worthy his reputed character for meekness and gentleness.

But I wanted to see the dove far from the "madding crowd" of blackbirds, blue jays, and red-heads, who, as well as himself, took corn for breakfast, and I set out to look him up. At first the whole family seemed to consist of the young, just flying about, sometimes accompanied by their mother. Apparently the fathers of the race were all off in the cooing business.

So early as the second of June I came upon my first pair of young doves, two charming little creatures, sitting placidly side by side. Grave, indeed, and very much grown-up looked these drab-coated little folk, silent and motionless, returning my gaze with an innocent openness that, it seemed to me, must disarm their most bitter enemy. When I came upon such a pair, as I frequently did, on the low branch of an apple-tree or a limb of their native cedar, I stopped instantly to look at them. Not an eyelid of the youngsters would move; if a head were turned as they heard me coming, it would remain at precisely that angle as long as I had patience to stay. They were invariably sitting down with the appearance of being prepared to stay all day, and almost always side by side, though looking in different directions, and one was always larger than the other. A lovely and picturesque group they never failed to make, and as for any show of hunger or impatience, one could hardly imagine they ever felt either. In every way they were a violent contrast to all their neighbors, the boisterous blue jays, lively catbirds, blustering robins, and the vulgar-mannered blackbirds.

Sometimes I chanced upon a mother sitting by her youngling, and although when I found her alone she always flew, beside her little charge she was dignified and calm in bearing, and looked at me with fearless eyes, relying, as it appeared, upon absolute stillness, and the resemblance of her color to the branches, to escape observation; a ruse which must generally be successful.

The nest, the remains of which I often saw on the tree where I found an infant, was the merest apology, hardly more than a platform, just enough to hold the pair of eggs which they are said always to contain. Indeed, no baby but a serene dove, with the repose of thirty generations behind it, could stay in it till his wings grew. As it is, he must be forced to perch, whether ready or not, for the structure cannot hold together long. The wonder is that the eggs do not roll out before they are hatched.

Several things made the bird an interesting subject for study; his reputation for meekness, his alleged silence,—except at wooing time,—and the halo of melancholy with which the poets have invested him. I resolved to make acquaintance with my gentle neighbor, and I sought and found a favorite retreat of the silent family. This was a grove away down in the southeast corner of the grounds, little visited by people, and beloved by birds of several kinds. Till June was half over, the high grass, that I could not bear to trample, prevented exploration in that direction, but as soon as it was cut I made a trip to the little grove, and found it a sort of doves' headquarters, and there, in many hours of daily study, I learned to know him a little, and respect him a good deal.

It was a delightful spot the doves had chosen to live in, and so frequented by birds that whichever way I turned my face, in two minutes I wished I had turned it the other, or that I had eyes in the back of my head. With reason, too, for the residents skipped around behind me, and all the interesting things went on at my back. I could hear the flit of wings, low, mysterious sounds, whispering, gentle complaints and hushings, but if I turned—lo! the scene shifted, and the drama of life was still enacted out of my sight. Yet I managed, in spite of this difficulty, to learn several things I did not know before.

No one attends to his own business more strictly than the dove. On the ground, where he came for corn, he seemed to see no other bird, and paid not the slightest heed to me in my window, but went about his own affairs in the most matter-of-fact way. Yet I cannot agree with the common opinion, which has made his name a synonym for all that is meek and gentle. He has a will of his own, and a "mild but firm" way of securing it. Sometimes, when all were busy at the corn, one of my Quaker-clad guests would take a notion, for what reason I could not discover, that some other dove must not stay, and he would drive him (or her) off. He was not rude or blustering, like the robin, nor did he make offensive remarks, after the manner of a blackbird; he simply signified his intention of having his neighbor go, and go he did, nolens volens.

It was droll to see how this "meek and gentle" fellow met blackbird impudence. If one of the sable gentry came down too near a dove, the latter gave a little hop and rustled his feathers, but did not move one step away. For some occult reason the blackbird seemed to respect this mild protest, and did not interfere again.

Would one suspect so solemn a personage of joking? yet what else could this little scene mean? A blackbird was on the ground eating, when a dove flew down and hovered over him as though about to alight upon him. It evidently impressed the blackbird exactly as it did me, for he scrambled out from under, very hastily. But the dove had no intention of the sort; he came calmly down on one side.

The first dove baby who accompanied its parent to the ground to be fed was the model of propriety one would expect from the demure infant already mentioned. He stood crouching to the ground in silence, fluttering his wings a little, but making no sound, either of begging, or when fed. A blackbird came to investigate this youngster, so different from his importunate offspring, upon which both doves flew.

There is a unique quality claimed for the dove: that with the exception of the well-known coo in nesting time he is absolutely silent, and that the noise which accompanies his flight is the result of a peculiar formation of the wing that causes a whistle. Of this I had strong doubts. I could not believe that a bird who has so much to say for himself during wooing and nesting time could be utterly silent the rest of the year; nor, indeed, do I believe that any living creature, so highly organized as the feathered tribes, can be entirely without expression.

I thought I would experiment a little, and one day, observing that a young dove spent most of his time alone on a certain cedar-tree, where a badly used-up nest showed that he had probably been hatched, or feeding on the ground near it, I resolved to see if I could draw him out. I passed him six times a day, going and coming from my meals, and I always stopped to look at him—a scrutiny which he bore unmoved, in dove fashion. So one morning, when I stood three feet from him, I began a very low whistle to him. He was at once interested, and after about three calls he answered me, very low, it is true, but still unmistakably. Though he replied, however, it appeared to make him uneasy, for while he had been in the habit of submitting to my staring without being in any way disconcerted, he now began to fidget about. He stood up, changed his place, flew to a higher branch, and in a few moments to the next tree; all the time, however, answering my calls.

I was greatly interested in my new acquaintance, and the next day I renewed my advances. As before, he answered, looking bright and eager, as I had never seen one of his kind look, and after three or four replies he became uneasy, as on the previous day, and in a moment he flew. But I was surprised and startled by his starting straight for me. I thought he would certainly alight on me, and such, I firmly believe, was his inclination, but he apparently did not quite dare trust me, so he passed over by a very few inches, and perched on the tree I was under. Then—still replying to me—he flew to the ground not six feet from me, and step by step, slowly moved away perhaps fifteen feet, when he turned and flew back to his own tree beside me. I was pleased to notice that the voice of this talkative dovekin was of the same quality as the "whistling" said to be of the wings, when a dove flies.

The last interview I had with the dear baby, I found him sitting with his back toward me, but the instant I whistled he turned around to face me, and seated himself again. He replied to me, and fluttered his wings slightly, yet he soon became restless, as usual. He did not fly, however, and he answered louder than he had done previously, but I found that my call must be just right to elicit a response. I might whistle all day and he would pay no attention, till I uttered a two-note call, the second note a third above the first and the two slurred together. I was delighted to find that even a dove, and a baby at that, could "talk back." He was unique in other ways; for example, in being content to pass his days in, and around, his own tree. I do not believe he had ever been farther than a small group of cedars, ten feet from his own. I always found him there, though he could fly perfectly well. This interview was, I regret to say, the last; the next morning my little friend was nowhere to be seen. Perhaps mamma thought he was getting too friendly with one of a race capable of eating a baby dove.

After this episode in my dove acquaintance, I was more than ever interested in getting at the mode of expression in the family, and I listened on every occasion. One day two doves alighted over my head when I was sitting perfectly still, and I distinctly heard very low talk, like that of my lost baby; there was, in addition, a note or two like the coo, but exceedingly low. I could not have heard a sound ten feet from the tree, nor if I had been stirring myself. I observed also that a dove can fly in perfect silence; and, moreover, that the whistle of the wings sometimes continues after the bird has become still. I heard the regular coo—the whole four-note performance —both in a whisper and in the ordinary tone, and the latter, though right over my head, sounded a mile away. At the end of my month's study I was convinced that the dove is far from being a silent bird; on the contrary, he is quite a talker, with the "low, sweet voice" so much desired in other quarters. And further, that the whistling is not produced wholly (if at all) by the wings, and it is a gross injustice to assert that he is not capable of expressing himself at all times and seasons.

BESIDE THE GREAT SALT LAKE

Up!—If thou knew'st who calls
To twilight parks of beach and pine
High o'er the river intervals,
Above the plowman's highest line,
Over the owner's farthest walls!
Up! where the airy citadel
O'erlooks the surging landscape's swell!

Emerson

XVIII. IN A PASTURE

The word "pasture," as used on the shore of the Great Salt Lake, conveys no true idea to one whose associations with that word have been formed in States east of the Rocky Mountains. Imagine an extensive inclosure on the side of a mountain, with its barren-looking soil strewn with rocks of all sizes, from a pebble to a bowlder, cut across by an irrigating ditch or a mountain brook, dotted here and there by sage bushes, and patches of oak-brush, and wild roses, and one has a picture of a Salt Lake pasture. Closely examined, it has other peculiarities. There is no half way in its growths, no shading off, so to speak, as elsewhere; not an isolated shrub, not a solitary tree, flourishes in the strange soil, but trees and shrubs crowd together as if for protection, and the clump, of whatever size or shape, ends abruptly, with the desert coming up to its very edge. Yet the soil, though it seems to be the driest and most unpromising of baked gray mud, needs nothing more than a little water, to clothe itself luxuriantly; the course of a brook or even an irrigating ditch, if permanent, is marked by a thick and varied border of greenery. What the poor creatures who wandered over those dreary wastes could find to eat was a problem to be solved only by close observation of their ways.

"H. H." said some years ago that the magnificent yucca, the glory of the Colorado mesas, was being exterminated by wandering cows, who ate the buds as soon as they appeared. The cattle of Utah—or their owners—have a like crime to answer for; not only do they constantly feed upon rose-buds and leaves, notwithstanding the thorns, but they regale themselves upon nearly every flower-plant that shows its head; lupines were the chosen dainty of my friend's horse. The animals become expert at getting this unnatural food; it is curious to watch the deftness with which a cow will go through a currant or gooseberry bush, thrusting her head far down among the branches, and carefully picking off the tender leaves, while leaving the stems untouched, and the matter-of-course way in which she will bend over and pull down a tall sapling, to despoil it of its foliage.

In a pasture such as I have described, on the western slope of one of the Rocky Mountains, desolate and forbidding though it looked, many hours of last summer's May and June "went their way," if not

"As softly as sweet dreams go down the night,"

certainly with interest and pleasure to two bird-students whose ways I have sometimes chronicled.

Most conspicuous, as we toiled upward toward our breezy pasture, was a bird whose chosen station was a fence—a wire fence at that. He was a tanager; not our brilliant beauty in scarlet and black, but one far more gorgeous and eccentric in costume, having, with the black wings and tail of our bird, a breast of shining yellow and a cap of crimson. His occupation on the sweet May mornings that he lingered with us, on his way up the mountains for the summer, was the familiar one of getting his living, and to that he gave his mind without reserve. Not once did he turn curious eyes upon us as we sauntered by or rested awhile to watch him. Eagerly his pretty head turned this way and that, but not for us; it was for the winged creatures of the air he looked, and when one that pleased his fancy fluttered by he dashed out and secured it, returning to a post or the fence just as absorbed and just as eager for the next one. Every time he alighted it was a few feet farther down the fence, and thus he worked his way out of our sight, without seeming aware of our existence.

This was not stupidity on the part of the crimson-head, nor was it foolhardiness; it was simply trust in his guardian, for he had one,—one who watched every movement of ours with close attention, whose vigilance was never relaxed, and who appeared, when we saw her, to be above the need of food. A plain personage she was, clad in modest, dull yellow,—the female tanager. She was probably his mate; at any rate, she gradually followed him down the fence, keeping fifteen or twenty feet behind him, all the time with an eye on us, ready to give warning of the slightest aggressive movement on our part. It would be interesting to know how my lord behaves up in those sky-parlors where their summer homes are made. No doubt he is as tender and devoted as most of his race (all his race, I would say, if Mr. Torrey had not shaken our faith in the ruby-throat), and I have no doubt that the little red-heads in the nest will be well looked after and fed by their fly-catching papa.

Far different from the cool unconcern of the crimson-headed tanager were the manners of another red-headed dweller on the mountain. The green-tailed towhee he is called in the books, though the red of his head is much more conspicuous than the green of his tail. In this bird the high-bred repose of his neighbor was replaced by the most fussy restlessness. When we surprised him on the lowest wire of the fence, he was terribly disconcerted, not to say thrown into a panic. He usually stood a moment, holding his long tail up in the air, flirted his wings, turned his body this way and that in great excitement, then hopped to the nearest bowlder, slipped down behind it, and ran off through the sage bushes like a mouse. More than this we were never able to see, and where he lived and how his spouse looked we do not know to this day.

Most interesting of the birds that we saw on our daily way to the pasture were the gulls; great, beautiful, snowy creatures, who looked strangely out of place so far away from the seashore. Stranger, too, than their change of residence was their change of manners from the wild, unapproachable sea-birds, soaring and diving, and apparently spending their lives on wings such as the poet sings,—

"When I had wings, my brother,
Such wings were mine as thine;"

and of whose lives he further says,—

"What place man may, we claim it,
But thine,—whose thought may name it?
Free birds live higher than freemen,
And gladlier ye than we."

From this high place in our thoughts, from this realm of
poetry and mystery, to come down almost to the tameness of the barnyard fowl is a marvelous transformation,
and one is tempted to believe the solemn announcement
of the Salt Lake prophet, that the Lord sent them to his
chosen people.

The occasion of this alleged special favor to the
Latter Day Saints was the advent, about twenty years
ago, of clouds of grasshoppers, before which the crops of
the Western States and Territories were destroyed as by
fire. It was then, in their hour of greatest need, when the
food upon which depended a whole people was threatened, that these beautiful winged messengers appeared.
In large flocks they came, from no one knows where, and
settled, like so many sparrows, all over the land, devour-
ing almost without ceasing the hosts of the foe. The
crops were saved, and all Deseret rejoiced. Was it any
wonder that a people trained to regard the head of their
church as the direct representative of the Highest should
believe these to be really birds of God, and should accordingly cherish them? Well would it be for themselves if
other Christian peoples were equally believing, and protected and cherished other winged messengers, sent just
as truly to protect their crops.

The shrewd man who wielded the destinies of his
people beside the Salt Lake secured the future usefulness
of what they considered the miraculous visitation by fixing a penalty of five dollars upon the head of every gull in

the Territory. And now, the birds having found congenial nesting-places on solitary islands in the lake, their descendants are so fearless and so tame that they habitually follow the plow like a flock of chickens, rising from almost under the feet of the indifferent horses and settling down at once in the furrow behind, seeking out and eating greedily all the worms and grubs and larvæ and mice and moles that the plow has disturbed in its passage. The Mormon cultivator has sense enough to appreciate such service, and no man or boy dreams of lifting a finger against his best friend.

Extraordinary indeed was this sight to eyes accustomed to seeing every bird who attempts to render like service shot and snared and swept from the face of the earth. Our hearts warmed toward the "Sons of Zion," and our respect for their intelligence increased, as we hurried down to the field to see this latter-day wonder.

Whether the birds distinguished between "saints" and sinners, or whether their confidence extended only to plow-boys, they would not let us come near them. But our glasses brought them close, and we had a very good study of them, finding exceeding interest in their ways: their quaint faces as they flew toward us; their dignified walk; their expression of disapproval, lifting the wings high above the back till they met; their queer and constant cries in the tone of a child who whines; and, above all, their use of the wonderful wings,—"half wing, half wave," Mrs. Spofford calls them.

To rise from the earth upon these beautiful great arms, seemed to be not so easy as it looks. Some of the graceful birds lifted them, and ran a little before leaving the ground, and all of them left both legs hanging, and both feet jerking awkwardly at every wing-beat, for a few moments on starting, before they carefully drew each flesh-colored foot up into its feather pillow,

> "And gray and silver up the dome
> Of gray and silver skies went sailing,"

in ever-widening circles, without moving a feather that we could perceive. It was charming to see how nicely they folded down their splendid wings on alighting, stretching each one out, and apparently straightening every feather before laying it into its place.

Several hours this interesting flock accompanied the horses and man around the field, taking possession of each furrow as it was laid open, and chattering and eating as fast as they could; and the question occurred to me, if a field that is thoroughly gleaned over every spring furnishes so great a supply of creatures hurtful to vegetation, what must be the state of grounds which are carefully protected from such gleaning, on which no bird is allowed to forage?

As noon approached, the hour when "birds their wise siesta take," although the plow did not cease its monotonous round, the birds retired in a body to the still untouched middle of the field, and settled themselves for their "nooning," dusting themselves—their snowy plumes!—like hens on an ash heap, sitting about in knots like parties of ducks, preening and shaking themselves out, or going at once to sleep, according to their several tastes. Half an hour's rest sufficed for the more active spirits, and then they treated us, their patient observers, to an aërial exhibition. A large number, perhaps three quarters of the flock, rose in a body and began a spiral flight. Higher and higher they went, in wider and wider circles, till, against the white clouds, they looked like a swarm of midges, and against the blue the eye could not distinguish them. Then from out of the sky dropped one after another, leaving the soaring flock looking wonderfully ethereal and gauzy in the clear air, with the sun above him, almost like a spirit bird gliding motionless through the ether till he alighted at last quietly beside his fellows on the ground. In another half hour they were all behind the plow again, hard at work.

When we had looked our fill, we straightway sought out and questioned some of the wise men among the "peculiar people." This is what we learned: that when plowing is over the birds retire to their home, an island in the lake, where, being eminently social birds, their nests are built in a community. Their beneficent service to mankind does not end with the plowing season, for when that is over they turn their attention to the fish that are brought into the lake by the fresh-water streams, at once strangled by its excess of salt, and their bodies washed up on the shore. What would become of the human residents if that animal deposit were left for the fierce sun to dispose of, may perhaps be imagined. The gull should, indeed, be a sacred bird in Utah.

What drew us first to the pasture—which we came to at last—was our search for a magpie's nest. The home of this knowing fellow is the Rocky Mountain region, and, naturally, he was the first bird we thought of looking for. There would be no difficulty in finding nests, we thought, for we came upon magpies everywhere in our walks. Now one alighted on a fence-post a few yards ahead of us, earnestly regarding our approach, tilting upward his long, expressive tail, the black of his plumage shining with brilliant blue reflections, and the white fairly dazzling the eyes. Again we caught glimpses of two or three of the beautiful birds walking about on the ground, holding their precious tails well up from the earth, and gleaning industriously the insect life of the horse pasture. At one moment we were saluted from the top of a tall tree, or shrieked at by one passing over our heads, looking like an immense dragonfly against the sky. Magpie voices were heard from morning till night; strange, loud calls of "mag! mag!" were ever in our ears. "Oh, yes," we had said, "we must surely go out some morning and find a nest."

First we inquired. Everybody knew where they built, in oak-brush or in apple-trees, but not a boy in that village knew where there was a nest. Oh, no, not one! A man confessed to the guilty secret, and, directed by him, we took a long walk through the village with its queer little houses, many of them having the two front doors which tell the tale of Mormondom within; up the long sidewalk, with a beautiful bounding mountain brook running down the gutter, as if it were a tame irrigating ditch, to a big gate in a "combination fence." What this latter might be we had wondered, but relied upon knowing it when we saw it,—and we did: it was a fence of laths held together by wires woven between them, and we recognized the fitness of the name instantly. Then on through the big gate, down a long lane where we ran the gauntlet of the family cows; over or under bars, where awaited us a tribe of colts with their anxious mammas; and at last to the tree and the nest. There our guide met us and climbed up to explore. Alas! the nest robber had anticipated us.

Slowly we took our way home, resolved to ask no more help, but to seek for ourselves, for the nest that is known is the nest that is robbed. So the next morning, armed with camp-chairs and alpenstocks, drinking-cups and notebooks, we started up the mountain, where we could at least find solitude, and the fresh air of the hills. We climbed till we were tired, and then, as was our custom, sat down to rest and breathe, and see who lived in that part of the world. Without thought of the height we had reached, we turned our backs to the mountain, rising bare and steep before us, and behold! the outlook struck us dumb.

There at our feet lay the village, smothered in orchards and shade-trees, the locusts, just then huge bouquets of graceful bloom and delicious odor, buzzing with hundreds of bees and humming-birds; beyond was a stretch of cultivated fields in various shades of green and brown; and then the lake,—beautiful and wonderful Salt Lake,—glowing with exquisite colors, now hyacinth blue, changing in places to tender green or golden brown, again sparkling like a vast bed of diamonds. In the foreground lay Antelope Island, in hues of purple and bronze, with its chain of hills and graceful sky-line; and resting on the horizon beyond were the peaks of the grand Oquirrhs, capped with snow. Well might we forget our quest while gazing on this impressive scene, trying to fix its various features in our memories, to be an eternal possession.

We were recalled to the business in hand by the sudden appearance on the top of a tree below us of one of the birds we sought. The branch bent and swayed as the heavy fellow settled upon it, and in a moment a comrade came, calling vigorously, and alighted on a neighboring branch. A few minutes they remained, with flirting tails, conversing in garrulous tones, then together they rose on broad wings, and passed away—away over the fields, almost out of sight, before they dropped into a patch of oak-brush. After them appeared others, and we sat there a long time, hoping to see at least one that had its home within our reach. But every bird that passed over turned its face to the mountains; some seemed to head for the dim Oquirrhs across the lake, while others disappeared over the top of the Wasatch behind us; not one paused in our neighborhood, excepting long enough to look at us, and express its opinion in loud and not very polite tones.

It was then and there that we noticed our pasture; the entrance was beside us. Shall we go in? was always the question before an inclosure. We looked over the wall. It was plainly the abode of horses, meek work-a-day beings, who certainly would not resent our intrusion. Oak-brush was there in plenty, and that is the chosen home of the magpie. We hesitated; we started for the gate. It was held in place by a rope elaborately and securely tied in many knots; but we had learned something about the gates of this "promised land,"—that between the posts and the stone wall may usually be found space enough to slip through without disturbing the fastenings.

In that country no one goes through a gate who can possibly go around it, and well is it indeed for the stranger and the wayfarer in "Zion" that such is the custom, for the idiosyncrasies of gates were endless; they agreed only in never fitting their place and never opening properly. If the gate was in one piece, it sagged so that it must be lifted; or it had lost one hinge, and fell over on the rash individual who loosened the fastenings; or it was about falling to pieces, and must be handled like a piece of choice bric-a-brac. If it had a latch, it was rusty or did not fit; and if it had not, it was fastened, either by a board slipped in to act as a bar and never known to be of proper size, or in some occult way which would require the skill of "the lady from Philadelphia" to undo. If it was of the fashion that opens in the middle, each individual gate had its particular "kink," which must be learned by the uninitiated before he—or, what is worse, she—could pass. Many were held together by a hoop or link of iron, dropped over the two end posts; but whether the gate must be pulled out or pushed in, and at exactly what angle it would consent to receive the link, was to be found out only by experience.

But not all gates were so simple even as this: the ingenuity with which a variety of fastenings,—all to avoid the natural and obvious one of a hook and staple,— had been evolved in the rural mind was fairly startling. The energy and thought that had been bestowed upon this little matter of avoiding a gate-hook would have built a bridge across Salt Lake, or tunneled the Uintas for an irrigating ditch.

Happily, we too had learned to "slip through," and we passed the gate with its rope puzzle, and the six or

eight horses who pointed inquiring ears toward their un-
wanted visitors, and hastened to get under cover before
the birds, if any lived there, should come home.

The oak-brush, which we then approached, is a
curious and interesting form of vegetation. It is a mass
of oak-trees, all of the same age, growing as close as they
can stand, with branches down to the ground. It looks as
if each patch had sprung from a great fall of acorns from
one tree, or perhaps were shoots from the roots of a per-
ished tree. The clumps are more or less irregularly
round, set down in a barren piece of ground, or among
the sage bushes. At a distance, on the side of a mountain,
they resemble patches of moss of varying shape. When
two or three feet high, one is a thick, solid mat; when it
reaches an altitude of six to eight feet, it is an impenetrable thicket; except, that is, when it happens to be in a
pasture. Horses and cattle find such scanty pickings in
the fields, that they nibble every green thing, even oak
leaves, and so they clear the brush as high as they can
reach. When therefore it is fifteen feet high, there is a
thick roof the animals are not able to reach, and one may
look through a patch to the light beyond. The stems and
lower branches, though kept bare of leaves, are so close
together and so intertwined and tangled, that forcing
one's way through it is an impossibility. But the horses
have made and kept open paths in every direction, and
this turns it into a delightful grove, a cool retreat, which
others appreciate as well as the makers.

Selecting a favorable-looking clump of oak-brush,
we attempted to get in without using the open horse
paths, where we should be in plain sight. Melancholy was
the result; hats pulled off, hair disheveled, garments

torn, feet tripped, and wounds and scratches innumerable. Several minutes of hard work and stubborn endurance enabled us to penetrate not more than half a dozen feet, when we managed, in some sort of fashion, to sit down, on opposite sides of the grove. Then, relying upon our "protective coloring" (not evolved, but carefully selected in the shops), we subsided into silence, hoping not to be observed when the birds came home, for there was the nest before us.

A wise and canny builder is Madam Mag, for though her home must be large to accommodate her size, and conspicuous because of the shallowness of the foliage above her, it is, in a way, a fortress, to despoil which the marauder must encounter a weapon not to be despised,—a stout beak, animated and impelled by indignant motherhood. The structure was made of sticks, and enormous in size; a half-bushel measure would hardly hold it. It was covered, as if to protect her, and it had two openings under the cover, toward either of which she could turn her face. It looked like a big, coarsely woven basket resting in a crotch up under the leaves, with a nearly close cover supported by a small branch above. The sitting bird could draw herself down out of sight, or she could defend herself and her brood, at either entrance.

In my retreat, I had noted all these points before any sign of life appeared in the brush. Then there came a low cry of "mag! mag!" and the bird entered near the ground. She alighted on a dead branch, which swung back and forth, while she kept her balance with her beautiful tail. She did not appear to look around; apparently she had no suspicions and did not notice us, sitting motionless and breathless in our respective places. Her head was turned to the nest, and by easy stages and with many pauses, she made her way to it. I could not see that she had a companion, for I dared not stir so much as a finger; but while she moved about near the nest there came to the eager listeners on the ground low and tender utterances in the sweetest of voices,—whether one or two I know not,—and at last a song, a true melody, of a yearning, thrilling quality that few song-birds, if any, can excel. I was astounded! Who would suspect the harsh-voiced, screaming magpie of such notes! I am certain that the bird or birds had no suspicion of listeners to the home talk and song, for after we were discovered, we heard nothing of the sort.

This little episode ended, madam slipped into her nest, and all became silent, she in her place and I in mine. If this state of things could only remain; if she would only accept me as a tree-trunk or a misshapen bowlder, and pay no attention to me, what a beautiful study I should have! Half an hour, perhaps more, passed without a sound, and then the silence was broken by magpie calls from without. The sitting bird left the nest and flew out of the grove, quite near the ground; I heard much talk and chatter in low tones outside, and they flew. I slipped out as quickly as possible, wishing indeed that I had wings as she had, and went home, encouraged to think I should really be able to study the magpie.

But I did not know my bird. The next day, before I knew she was about, she discovered me, though it was plain that she hoped I had not discovered her. Instantly she became silent and wary, coming to her nest over the top of the trees, so quietly that I should not have known it except for her shadow on the leaves. No talk or song now fell upon my ear; calls outside were few and subdued. Everything was different from the natural unconsciousness of the previous day; the birds were on guard, and henceforth I should be under surveillance.

From this moment I lost my pleasure in the study, for I feel little interest in the actions of a bird under the constraint of an unwelcome presence, or in the shadow of constant fear and dread. What I care to see is the natural life, the free, unstudied ways of birds who do not notice or are not disturbed by spectators. Nor have I any pleasure in going about the country staring into every tree, and poking into every bush, thrusting irreverent hands into the mysteries of other lives, and rudely tearing away the veils that others have drawn around their private affairs. That they are only birds does not signify to me; for me they are fellow-creatures; they have rights, which I am bound to respect.

I prefer to make myself so little obvious, or so apparently harmless to a bird, that she will herself show me her nest, or at least the leafy screen behind which it is hidden. Then, if I take advantage of her absence to spy upon her treasures, it is as a friend only,—a friend who respects her desire for seclusion, who never lays profane hands upon them, and who shares the secret only with one equally reverent and loving. Naturally I do not find so many nests as do the vandals to whom nothing is sacred, but I enjoy what I do find, in a way it hath not entered into their hearts to conceive.

In spite of my disinclination, we made one more call upon the magpie family, and this time we had a reception. This bird is intelligent and by no means a slave to habit; because he has behaved in a certain way once, there is no law, avian or divine, that compels him to repeat that conduct on the next occasion. Nor is it safe to generalize about him, or any other bird for that matter. One cannot say, "The magpie does thus and so," because each individual magpie has his own way of doing, and circumstances alter cases, with birds as well as with people.

On this occasion we placed ourselves boldly, though very quietly, in the paths that run through the oak-brush. We had abandoned all attempt at concealment; we could hope only for tolerance. The birds readily understood; they appreciated that they were seen and watched, and their manners changed accordingly. The first one of the black-and-white gentry who entered the grove discovered my comrade, and announced the presence of the enemy by a loud cry, in what somebody has aptly called a "frontier tone of voice." Instantly another appeared and added his remarks; then another, and still another, till within five minutes there were ten or twelve excited magpies, shouting at the top of their voices, and hopping and flying about her head, coming ever nearer and nearer, as if they meditated a personal attack. I did not really fear it, but I kept close watch, while remaining motionless, in the hope that they would not notice me. Vain hope! nothing could escape those sharp eyes when once the bird was aroused. After they had said what they chose to my friend, who received the taunts and abuse of the infuriated mob in meek silence, lifting not her voice to reply, they turned the stream of their eloquence upon me.

I was equally passive, for indeed I felt that they had a grievance. We have no right to expect birds to tell one human being from another, so long as we, with all our boasted intelligence, cannot tell one crow or one magpie from another; and all the week they had suffered persecution at the hands of the village boys. Young magpies, nestlings, were in nearly every house, and the birds had endured pillage, and doubtless some of them death. I did not blame the grieved parents for the reception they gave us; from their point of view we belonged to the enemy.

After the storm had swept by, and while we sat there waiting to see if the birds would return, one of the horses of the pasture made his appearance on the side where I sat, now eating the top of a rosebush, now snipping off a flower plant that had succeeded in getting two leaves above the ground, but at every step coming nearer me. It was plain that he contemplated retiring to this shady grove, and, not so observing as the magpies, did not see that it was already occupied. When he was not more than ten feet away, I snatched off my sun hat and waved it before him, not wishing to make a noise. He stopped instantly, stared wildly for a moment, as if he had never seen such an apparition, then wheeled with a snort, flung out his heels in disrespect, and galloped off down the field.

The incident was insignificant, but the result was curious. So long as we stayed in that bit of brush, not a horse attempted to enter, though they all browsed around outside. They avoided it as if it were haunted, or, as my comrade said, "filled with beckoning forms." Nor was that all; I have reason to think they never again entered that particular patch of brush, for, some weeks after we had abandoned the study of magpies and the pasture altogether, we found the spot transformed, as if by the wand of enchantment. From the burned-up desert outside we stepped at once into a miniature paradise, to our surprise, almost our consternation. Excepting the footpaths through it, it bore no appearance of having ever been a thoroughfare. Around the foot of every tree had grown up clumps of ferns or brakes, a yard high, luxuriant, graceful, and exquisite in form and color; and peeping out from under them were flowers, dainty wildings we had not before seen there. A bit of the tropics or a gem out of fairyland it looked to our sun and sand weary eyes. Outside were the burning sun of June, a withering hot wind, and yellow and dead vegetation; within was cool greenness and a mere rustle of leaves whispering of the gale. It was the loveliest bit of greenery we saw on the shores of the Great Salt Lake. It was marvelous; it was almost uncanny.

Our daily trips to the pasture had ceased, and other birds and other nests had occupied our thoughts for a week or two, when we resolved to pay a last visit to our old haunts, to see if we could learn anything of the magpies. We went through the pasture, led by the voices of the birds away over to the farther side, and there, across another fenced pasture, we heard them plainly, calling and chattering and making much noise, but in different tones from any we had heard before. Evidently a magpie nursery had been established over there. We fancied we could distinguish maternal reproof and loving baby talk, beside the weaker voices of the young, and we went home rejoicing to believe, that in spite of nest robbers, and the fright we had given them, some young magpies were growing up to enliven the world another summer.

XIX. THE SECRET OF THE WILD ROSE PATH

"Shall I call thee Bird,
Or but a wandering Voice?"

Wordsworth's lines are addressed to the cuckoo of the Old World, a bird of unenviable reputation, notorious for imposing his most sacred duties upon others; naturally, therefore, one who would not court observation, and whose ways would be somewhat mysterious. But the American representative of the family is a bird of different manners. Unlike his namesake across the water, our cuckoo never—or so rarely as practically to be never—shirks the labor of nest-building and raising a family. He has no reason to skulk, and though always a shy bird, he is no more so than several others, and in no sense is he a mystery.

There is, however, one American bird for whom Wordsworth's verse might have been written; one whose chief aim seems to be, reversing our grandmothers' rule for little people, to be heard, and not seen. To be seen is, with this peculiar fellow, a misfortune, an accident, which he avoids with great care, while his voice rings out loud and clear above all others in the shrubbery. I refer to the yellow-breasted chat (Icteria virens), whose summer home is the warmer temperate regions of our country, from the Atlantic to the Pacific coast, and whose unbird-like utterances prepare one to believe the stories told of his eccentric actions; this, for example, by Dr. Abbott:—

> "Aloft in the sunny air he springs;
> To his timid mate he calls;
> With dangling legs and fluttering wings
> On the tangled smilax falls;
> He mutters, he shrieks—
> A hopeless cry;
> You think that he seeks
> In peace to die,
> But pity him not; 't is the ghostly chat,
> An imp if there is one, be sure of that."

I first knew the chat—if one may be said to know a creature so shy—in a spot I have elsewhere described, a deserted park at the foot of Cheyenne Mountain. I became familiar with his various calls and cries (one can hardly call them songs); I secured one or two fleeting glimpses of his graceful form; I sought and discovered the nest, which thereupon my Lady Chat promptly abandoned, though I had not laid a finger upon it; and last of all, I had the sorrow and shame of knowing that my curiosity had driven the pair from the neighborhood. This was the Western form of Icteria, differing from the Eastern only in a greater length of tail, which several of our Rocky Mountain birds affect, for the purpose, apparently, of puzzling the ornithologist.

Two years after my unsuccessful attempt to cultivate friendly relations with "the ghostly chat," the middle of May found me on the shore of the Great Salt Lake, where I settled myself at the foot of the Wasatch Mountains, at that point bare, gray, and unattractive, showing miles of loose bowlders and great patches of sage-bush. In the monotonous stretches of this shrub, each plant of which looks exactly like every other, dwelt many shy birds, as well hidden as bobolinks in the meadow grass, or meadow-larks in the alfalfa.

But on this mountain side no friendly cover existed from which I could spy out bird secrets. Whatever my position, and wherever I placed myself, I was as conspicuous as a tower in the middle of a plain; again, no shadow of protection was there from the too-ardent sun of Utah, which drew the vitality from my frame as it did the color from my gown; worse than these, the everywhere present rocks were the chosen haunts of the one enemy of a peaceful bird lover, the rattlesnake, and I hesitated to pursue the bird, because I invariably forgot to watch and listen for the reptile. Bird study under these conditions was impossible, but the place presented a phase of nature unfamiliar to me, and for a time so fascinating that every morning my steps turned of themselves "up the stony pathway to the hills."

The companion of my walks, a fellow student of birds, was more than fascinated; she was enraptured. The odorous bush had associations for her; she reveled in it; she inhaled its fragrance as a delicious perfume; she filled her pockets with it; she lay for hours at a time on the ground, where she could bask in the sunshine, and see nothing but the gray leaves around her and the blue sky above.

I can hardly tell what was the fascination for me. It was certainly not the view of the mountains, though mountains are beyond words in my affections. The truth is, the Rocky Mountains, many of them, need a certain distance to make them either picturesque or dignified. The range then daily before our eyes, the Wasatch, was, to dwellers at its feet, bleak, monotonous, and hopelessly prosaic. The lowest foothills, being near, hid the taller peaks, as a penny before the eye will hide a whole landscape.

Let me not, however, be unjust to the mountains I love. There is a range which satisfies my soul, and will rest in my memory forever, a beautiful picture, or rather a whole gallery of pictures. I can shut my eyes and see it at this moment, as I have seen it a thousand times. In the early morning, when the level sun shines on its face, it is like one continuous mountain reaching across the whole western horizon; it has a broken and beautiful sky line; Pike's Peak looms up toward the middle, and lovely Cheyenne ends it in graceful slope on the south; lights and shadows play over it; its colors change with the changing sky or atmosphere,—sometimes blue as the heavens, sometimes misty as a dream; it is wonderfully beautiful then. But wait till the sun gets higher; look again at noon, or a little later. Behold the whole range has sprung into life, separated into individuals; gorges are cut where none had appeared; chasms come to light; cañons and all sorts of divisions are seen; foothills move forward to their proper places, and taller peaks turn at angles to each other; shapes and colors that one never suspected come out in the picture: the transformation is marvelous. But the sun moves on, the magical moment passes, each mountain slips back into line, and behold, you see again the morning's picture.

Indulge me one moment, while I try to show you the last picture impressed upon my memory as the train bore me, unwilling, away. It was cloudy, a storm was coming up, and the whole range was in deep shadow, when suddenly through some rift in the clouds a burst of sunshine fell upon the "beloved mountain" Cheyenne, and upon it alone. In a moment it was a smiling picture,

"Glad
With light as with a garment it was clad;"

all its inequalities, its divisions, its irregularities emphasized, its greens turned greener, its reds made more glowing,—an unequaled gem for a parting gift.

To come back to Utah. One morning, on our way up to the heights, as we were passing a clump of oakbrush, a bird cry rang out. The voice was loud and clear, and the notes were of a peculiar character: first a "chack" two or three times repeated, then subdued barks like those of a distressed puppy, followed by hoarse "mews" and other sounds suggesting almost any creature rather than one in feathers. But with delight I recognized the chat; my enthusiasm instantly revived. I unfolded my camp chair, placed myself against a stone wall on the opposite side of the road, and became silent and motionless as the wall itself.

My comrade, on the contrary, as was her custom, proceeded with equal promptness to follow the bird up, to hunt him out. She slipped between the barbed wires which, quite unnecessarily, one would suppose, defended the bleak pasture from outside encroachment, and passed out of sight down an obscure path that led into the brush where the bird was hidden. Though our ways differ, or rather, perhaps, because our ways differ, we are able to study in company. Certainly this circumstance proved available in circumventing the wily chat, and that happened which had happened before: in fleeing from one who made herself obvious to him, he presented himself, an unsuspecting victim, to another who sat like a statue against the wall. To avoid his pursuer, the bird slipped through the thick foliage of the low oaks, and took his place on the outside, in full view of me, but looking through the branches at the movements within so intently that he never turned his eyes toward me. This gave me an opportunity to study his manners that is rare indeed, for a chat off his guard is something ordinarily inconceivable.

He shouted out his whole répertoire (or so it seemed) with great vehemence, now "peeping" like a bird in the nest, then "chacking" like a blackbird, mewing as neatly as pussy herself, and varying these calls by the rattling of castanets and other indescribable sounds. His perch was half way down the bush; his trim olive-drab back and shining golden breast were in their spring glory, and he stood nearly upright as he sang, every moment stretching up to look for the invader behind the leaves. The instant she appeared outside, he vanished

within, and I folded my chair and passed on. His distur-
ber had not caught a glimpse of him.

My next interview with a chat took place a day or
two later. Between the cottage which was our temporary
home and the next one was a narrow garden bordered by
thick hedges, raspberry bushes down each side, and a
mass of flowering shrubs next the street. From my seat
within the house, a little back from the open window, I
was startled by the voice of a chat close at hand. Looking
cautiously out, I saw him in the garden, foraging about
under cover of the bushes, near the ground, and there for
some time I watched him. He had not the slightest repose of manner; the most ill-bred tramp in the English
sparrow family was in that respect his superior, and the
most nervous and excitable of wrens could not outdo
him in posturing, jerking himself up, flirting his tail, and
hopping from twig to twig. When musically inclined, he
perched on the inner side of the bushes against the front
fence, a foot or two above the ground, and within three
feet of any one who might pass, but perfectly hidden.

The performance of the chat was exceedingly
droll; first a whistle, clear as an oriole note, followed by
chacks that would deceive a red-wing himself, and then,
oddest of all, the laugh of a feeble old man, a weak sort of
"yah! yah! yah!" If I had not seen him in the act, I could
not have believed the sound came from a bird's throat.
He concluded with a low, almost whispered "chur-r-r," a
sort of private chuckle over his unique exhibition. After a
few minutes' singing he returned to his foraging on the
ground, or over the lowest twigs of the bushes, all the
time bubbling over with low joyous notes, his graceful
head thrown up, and his beautiful golden throat swelling

with the happy song. The listener and looker behind the screen was charmed to absolute quiet, and the bird so utterly unsuspicious of observers that he was perfectly natural and at his ease, hopping quickly from place to place, and apparently snatching his repast between notes.

The chat's secret of invisibility was thus plainly revealed. It is not in his protective coloring, for though his back is modest of hue, his breast is conspicuously showy; nor is it in his size, for he is almost as large as an oriole; it is in his manners. The bird I was watching never approached the top of a shrub, but invariably perched a foot or more below it, and his movements, though quick, were silence itself. No rustle of leaves proclaimed his presence; indeed, he seemed to avoid leaves, using the outside twigs near the main stalk or trunk, where they are usually quite bare, and no flit of wing or tail gave warning of his change of position. There was a seemingly natural wariness and cautiousness in every movement and attitude, that I never saw equaled in feathers.

Then, too, the clever fellow was so constantly on his guard and so alert that the least stir attracted his attention. Though inside the house, as I said, not near the window, and further veiled by screens, I had to remain as nearly motionless as possible, and use my glass with utmost caution. The smallest movement sent him into the bushes like a shot,—or rather, like a shadow, for the passage was always noiseless. Suspicion once aroused, the bird simply disappeared. One could not say of him, as of others, that he flew, for whether he used his wings, or melted away, or sank into the earth, it would be hard to tell. All I can be positive about is, that whereas one moment he was there, the next he was gone.

After this exhibition of the character of the chat, his constant watchfulness, his distrust, his love of mystery, it may appear strange that I should try again to study him at home, to find his nest and see his family. But there is something so bewitching in his individuality, that, though I may be always baffled, I shall never be discouraged. Somewhat later, when it was evident that his spouse had arrived and domestic life had begun, and I became accustomed to hearing a chat in a certain place every day as I passed, I resolved to make one more effort to win his confidence, or, if not that, at least to win his tolerance.

The chat medley for which I was always listening came invariably from one spot on my pathway up the mountain. It was the lower end of a large horse pasture, and near the entrance stood a small brick house, in which no doubt dwelt the owner, or care-taker, of the animals. The wide gate, in a common fashion of that country, opened in the middle, and was fastened by a link of iron which dropped over the two centre posts. The rattle of the iron as I touched it, on the morning I resolved to go in, brought to the door a woman. She was rather young, with hair cut close to her head, and wore a dark cotton gown, which was short and scant of skirt, and covered with a "checked apron." She was evidently at work, and was probably the mistress, since few in that "working-bee" village kept maids.

I made my request to go into the pasture to look at the birds.

"Why, certainly," she said, with a courtesy that I have found everywhere in Utah, though with a slow surprise growing in her face. "Come right in."

I closed and fastened the gate, and started on past her. Three feet beyond the doorsteps I was brought to a standstill: the ground as far as I could see was water-soaked; it was like a saturated sponge. Utah is dominated by Irrigation; she is a slave to her water supply. One going there from the land of rains has much to learn of the possibilities and the inconveniences of water. I was always stumbling upon it in new combinations and unaccustomed places, and I never could get used to its vagaries. Books written in the interest of the Territory indulge in rhapsodies over the fact that every man is his own rain-maker; and I admit that the arrangement has its advantages—to the cultivator. But judging from the standpoint of an outsider, I should say that man is not an improvement upon the original providence which distributes the staff of life to plants elsewhere, spreading the vital fluid over the whole land, so evenly that every grass blade gets its due share; and as all parts are wet at once, so all are dry at the same time, and the surplus, if there be any, runs in well-appointed ways, with delight to both eye and ear. All this is changed when the office of Jupiter Pluvius devolves upon man; different indeed are his methods. A man turns a stream loose in a field or pasture, and it wanders whither it will over the ground. The grass hides it, and the walker, bird-student or botanist, steps splash into it without the slightest warning. This is always unpleasant, and is sometimes disastrous, as when one attempts to cross the edge of a field of some close-growing crop, and instantly sinks to the top of the shoes in the soft mud.

On the morning spoken of, I stopped before the barrier, considering how I should pass it, when the woman showed me a narrow passage between the house and the stone wall, through which I could reach the higher ground at the back. I took this path, and in a moment was in the grove of young oaks which made her out-of-doors kitchen and yard. A fire was burning merrily in the stove, which stood under a tree; frying-pans and baking-tins, dippers and dishcloths, hung on the outer wall of her little house, and the whole had a camping-out air that was captivating, and possible only in a rainless land. I longed to linger and study this open-air housekeeping; if that woman had only been a bird!

But I passed on through the oak-grove back yard, following a path the horses had made, till I reached an open place where I could overlook the lower land, filled with clumps of willows with their feet in the water, and rosebushes

"O'erburdened with their weight of flowers,
And drooping 'neath their own sweet scent."

A bird was singing as I took my seat, a grosbeak,—perhaps the one who had entertained me in the field below, while I had waited hour after hour, for his calm-eyed mate to point out her nest. He sang there from the top of a tall tree, and she busied herself in the low bushes, but up to that time they had kept their secret well. He was a beautiful bird, in black and orange-brown and gold,—the black-headed grosbeak; and his song, besides being very pleasing, was interesting because it seemed hard to get out. It was as if he had conceived a brilliant and beautiful strain, and found himself unable to execute it. But if he felt the incompleteness of his performance as I did, he did not let it put an end to his endeavor. I sat there listening, and he came nearer, even to a low tree over my head; and as I had a glimpse or two of his mate in a tangle of willow and roses far out in the wet land, I concluded he was singing to her, and not to me. Now that he was so near, I heard more than I had before, certain low, sweet notes, plainly not intended for the public ear. This undertone song ended always in "sweet! sweet! sweet!" usually followed by a trill, and was far more effective than his state performances. Sometimes, after the "sweet" repeated half a dozen times, each note lower than the preceding one, he ended with a sort of purr of contentment.

I became so absorbed in listening that I had almost forgotten the object of my search, but I was suddenly recalled by a loud voice at one side, and the lively genius of the place was on hand in his usual rôle. Indeed, he rather surpassed himself in mocking and taunting cries that morning, either because he wished, as my host, to entertain me, or, what was more probable, to reproach me for disturbing the serenity of his life. Whatever might have been his motive, he delighted me, as always, by the spirit and vigor with which he poured out his chacks and whistles and rattles and calls. Then I tried to locate him by following up the sound, picking my way through the bushes, and among the straggling arms of the irrigating stream. After some experiments, I discovered that he was most concerned when I came near an impenetrable tangle that skirted the lower end of the lot. I say "near:" it was near "as the crow flies," but for one without wings it may have been half a mile; for between me and that spot was a great gulf fixed, the rallying point of the most erratic of wandering streamlets, and so given over to its vagaries that no bird-gazer, however enthusiastic, and indifferent to wet feet and draggled garments, dared attempt to pass. There I was forced to pause, while the bird flung out his notes as if in defiance, wilder, louder, and more vehement than ever.

In that thicket, I said to myself, as I took my way home, behind that tangle, if I can manage to reach it, I shall find the home of the chat. The situation was discouraging, but I was not to be discouraged; to reach that stronghold I was resolved, if I had to dam up the irrigator, build a bridge, or fill up the quagmire.

No such heroic treatment of the difficulty was demanded; my problem was very simply solved. As I entered the gate the next morning, my eyes fell upon an obscure footpath leading away from the house and the watery way beyond it, down through overhanging wild roses, and under the great tangle in which the chat had hidden. It looked mysterious, not to say forbidding, and, from the low drooping of the foliage above, it was plainly a horse path, not a human way. But it was undoubtedly the key to the secrets of the tangle, and I turned into it without hesitation. Stooping under the branches hanging low with their fragrant burden, and stopping every moment to loosen the hold of some hindering thorn, I followed in the footsteps of my four-footed pioneers till I reached the lower end of the marsh that had kept me from entering on the upper side. On its edge I placed my chair and seated myself.

It was an ideal retreat; within call if help were needed, yet a solitude it was plain no human being, in that land where (according to the Prophet) every man, woman, and child is a working bee, ever invaded;

"A leafy nook
Where wind never entered, nor branch ever shook,"

known only to my equine friends and to me. I exulted in it! No discoverer of a new land, no stumbler upon a gold mine, was ever more exhilarated over his find than I over my solitary wild rose path.

The tangle was composed of a varied growth. There seemed to have been originally a straggling row of low trees, chokecherry, peach, and willow, which had been surrounded, overwhelmed, and almost buried by a rich growth of shoots from their own roots, bound and cemented together by the luxuriant wild rose of the West, which grows profusely everywhere it can get a foothold, stealing up around and between the branches, till it overtops and fairly smothers in blossoms a fair-sized oak or other tree. Besides these were great ferns, or brakes, three or four feet high, which filled up the edges of the thicket, making it absolutely impervious to the eye, as well as to the foot of any straggler. Except in the obscure passages the horses kept open, no person could penetrate my jungle.

I had hardly placed myself, and I had not noted half of these details, when it became evident that my presence disturbed somebody. A chat cried out excitedly, "chack! chack! whe-e-w!" whereupon there followed an angry squawk, so loud and so near that it startled me. I turned quickly, and saw madam herself, all ruffled as if from the nest. She was plainly as much startled as I was, but she scorned to flee. She perked up her tail till she looked like an exaggerated wren; she humped her shoulders; she turned this way and that, showing in every movement her anger at my intrusion; above all, she repeated at short intervals that squawk, like an enraged hen. Hearing a rustle of wings on the other side, I turned my eyes an instant, and when I looked again she had gone! She would not run while I looked at her, but she had the true chat instinct of keeping out of sight.

She did not desert her grove, however. The canopy over my head, the roof to my retreat, was of green leaves, translucent, almost transparent. The sun was the sun of Utah; it cast strong shadows, and not a bird could move without my seeing it. I could see that she remained on guard, hopping and flying silently from one point of view to another, no doubt keeping close watch of me all the time.

Meanwhile the chat himself had not for a moment ceased calling. For some time his voice would sound quite near; then it would draw off, growing more and more distant, as if he were tired of watching one who did absolutely nothing. But he never got far away before madam recalled him, sometimes by the squawk alone, sometimes preceding it by a single clear whistle, exactly in his own tone. At once, as if this were a signal, —which doubtless it was,—his cries redoubled in energy, and seemed to come nearer again.

Above the restless demonstrations of the chats I could hear the clear, sweet song of the Western meadowlark in the next field. Well indeed might his song be serene; the minstrel of the meadow knew perfectly well that his nest and nestlings were as safely hidden in the middle of the growing lucern as if in another planet; while the chat, on the contrary, was plainly conscious of the ease with which his homestead might be discovered. A ruthless destroyer, a nest-robbing boy, would have had the whole thing in his pocket days ago. Even I, if I had not preferred to have the owners show it to me: if I had not made excuses to myself, of the marsh, of bushes too low to go under; if I had not hated to take it by force, to frighten the little folk I wished to make friends with,— even I might have seen the nest long before that morning. Thus I meditated as, after waiting an hour or two, I started for home.

Outside the gate I met my fellow-student, and we went on together. Our way lay beside an old orchard that we had often noticed in our walks. The trees were not far apart, and so overgrown that they formed a deep shade, like a heavy forest, which was most attractive when everything outside was baking in the June sun. It was nearly noon when we reached the gate, and looking into a place

"So curtained with trunks and boughs
That in hours when the ringdove coos to his spouse
The sun to its heart scarce a way could win,"

we could not resist its inviting coolness; we went in.

As soon as we were quiet, we noticed that there were more robins than we had heretofore seen in one neighborhood in that part of the world; for our familiar bird is by no means plentiful in the Rocky Mountain countries, where grassy lawns are rare, and his chosen food is not forthcoming. The old apple-trees seemed to be a favorite nesting-place, and before we had been there five minutes we saw that there were at least two nests within fifty feet of us, and a grosbeak singing his love song, so near that we had hopes of finding his home, also, in this secluded nook.

The alighting of a bird low down on the trunk of a tree, perhaps twenty feet away, called the attention of my friend to a neighbor we had not counted upon, a large snake, with, as we noted with horror, the color and markings of the dreaded rattler. He had, as it seemed, started to climb one of the leaning trunks, and when he had reached a point where the trunk divided into two parts, his head about two feet up, and the lower part of his body still on the ground, had stopped, and now rested thus, motionless as the tree itself. It may be that it was the sudden presence of his hereditary enemy that held him apparently spellbound, or it is possible that this position served his own purposes better than any other. Our first impulse was to leave his lordship in undisputed possession of his shady retreat; but the second thought, which held us, was to see what sort of reception the robins would give him. There was a nest full of young on a neighboring tree, and it was the mother who had come down to interview the foe. Would she call her mate? Would the neighbors come to the rescue? Should we see a fight, such as we had read of? We decided to wait for the result.

Strange to say, however, this little mother did not call for help. Not one of the loud, disturbed cries with which robins greet an innocent bird-student or a passing sparrow hawk was heard from her; though her kinsfolk sprinkled the orchard, she uttered not a sound. For a moment she seemed dazed; she stood motionless, staring at the invader as if uncertain whether he were alive. Then she appeared to be interested; she came a little nearer, still gazing into the face of her enemy, whose erect head and glittering eyes were turned toward her. We could not see that he made the slightest movement, while she hopped nearer and nearer; sometimes on one division of the trunk, and sometimes on the other, but always, with every hop, coming a little nearer. She did not act fright-ened nor at all anxious; she simply seemed interested, and inclined to close investigation. Was she fascinated? Were the old stories of snake power over birds true? Our interest was most intense; we did not take our eyes from her; nothing could have dragged us away then.

Suddenly the bird flew to the ground, and, so quickly that we did not see the movement, the head of the snake was turned over toward her, proving that it was the bird, and not us, he was watching. Still she kept drawing nearer till she was not more than a foot from him, when our sympathy with the unfortunate creature, who apparently was unable to tear herself away, overcame our scientific curiosity. "Poor thing, she'll be killed! Let us drive her away!" we cried. We picked up small stones which we threw toward her; we threatened her with sticks; we "shooed" at her with demonstrations that would have quickly driven away a robin in possession of its senses. Not a step farther off did she move; she hopped one side to avoid our missiles, but instantly fluttered back to her doom. Meanwhile her mate appeared upon the scene, hovering anxiously about in the trees overhead, but not coming near the snake.

By this time we had lost all interest in the question whether a snake can charm a bird to its destruction; we thought only of saving the little life in such danger. We looked around for help; my friend ran across the street to a house, hurriedly secured the help of a man with a heavy stick, and in two minutes the snake lay dead on the ground.

The bird, at once relieved, flew hastily to her nest, showing no signs of mental aberration, or any other effect of the strain she had been under. The snake was what the man called a "bull snake," and so closely resembled the rattler in color and markings that, although its exterminator had killed many of the more famous reptiles, he could not tell, until it was stretched out in death, which of the two it was. This tragedy spoiled the old orchard for me, and never again did I enter its gates.

Down the wild rose path I took my way the next morning. Silently and quickly I gained my seat of yesterday, hoping to surprise the chat family. No doubt my hope was vain; noiseless, indeed, and deft of movement must be the human being who could come upon this alert bird unawares. He greeted me with a new note, a single clear call, like "ho!" Then he proceeded to study me, coming cautiously nearer and nearer, as I could see out of the corner of my eye, while pretending to be closely occupied with my notebook. His loud notes had ceased, but it is not in chat nature to be utterly silent; many low sounds dropped from his beak as he approached. Sometimes it was a squawk, a gentle imitation of that which rang through the air from the mouth of his spouse; again it was a hoarse sort of mewing, followed by various indescribable sounds in the same undertone;

and then he would suddenly take himself in hand, and be perfectly silent for half a minute.

After a little, madam took up the matter, uttering her angry squawk, and breaking upon my silence almost like a pistol shot. At once I forgot her mate, and though he retired to a little distance and resumed his brilliant musical performance, I did not turn my head at his beguilements. She was the business partner of the firm whose movements I wished to follow. She must, sooner or later, go to her nest, while he might deceive me for days. Indeed, I strongly suspected him of that very thing, and whenever he became bolder in approaching, or louder and more vociferous of tongue, I was convinced that it was to cover her operations. I redoubled my vigilance in watching for her, keeping my eyes open for any slight stirring of a twig, tremble of a leaf, or quick shadow near the ground that should point her out as she skulked to her nest. I had already observed that whenever she uttered her squawks he instantly burst into energetic shouts and calls. I believed it a concerted action, with the intent of drawing my attention from her movements.

On this day the disturbed little mother herself interviewed me. First she came silently under the green canopy, in plain sight, stood a moment before me, jerking up her beautiful long tail and letting it drop slowly back, and posing her mobile body in different positions; then suddenly flying close past me, she alighted on one side, and stared at me for half a dozen seconds. Then, evidently, she resolved to take me in hand. She assumed the rôle of deceiver, with all the wariness of her family; her object being, as I suppose, carefully to point out where her nest was not. She circled about me, taking no pains to avoid my gaze. Now she squawked on the right; then she acted "the anxious mother" on the left; this time it was from the clump of rosebushes in front that she rose hurriedly, as if that was her home; again it was from over my head, in the chokecherry-tree, that she bustled off, as if she had been "caught in the act." It was a brilliant, a wonderful performance, a thousand times more effective than trailing or any of the similar devices by which an uneasy bird mother draws attention from her brood. It was so well done that at each separate manœuvre I could hardly be convinced by my own eyes that the particular spot indicated did not conceal the little homestead I was seeking. Several times I rose triumphant, feeling sure that "now indeed I do know where it is," and proceeded at once to the bush she had pointed out with so much simulated reluctance, parted the branches, and looked in, only to find myself deceived again. Her acting was marvelous. With just the properly anxious, uneasy manner, she would steal behind a clump of leaves into some retired spot admirably adapted for a chat's nest, and after a moment sneak out at the other side, and fly away near the ground, exactly as all bird-students have seen bird mothers do a thousand times.

After this performance a silence fell upon the tangle and the solitary nook in which I sat,—and I meditated. It was the last day of my stay. Should I set up a search for that nest which I was sure was within reach? I could go over the whole in half an hour, examine every shrub and low tree and inch of ground in it, and doubtless I should find it. No; I do not care for a nest thus forced. The distress of parents, the panic of nestlings, give me no pleasure. I know how a chat's nest looks. I have seen one with its pinky-pearl eggs; why should I care to see another? I know how young birds look; I have seen dozens of them this very summer. Far better that I never lay eyes upon the nest than to do it at such cost.

As I reached this conclusion, into the midst of my silence came the steady tramp of a horse. I knew the wild rose path was a favorite retreat from the sun, and it was very hot. The path was narrow; if a horse came in upon me, he could not turn round and retreat, nor was there room for him to pass me. Realizing all this in an instant, I snatched up my belongings, and hurried to get out before he should get in.

When I emerged, the chat set up his loudest and most triumphant shouts. "Again we have fooled you," he seemed to say; "again we have thrown your poor human acuteness off the scent! We shall manage to bring up our babies in safety, in spite of you!"

So indeed they might, even if I had seen them; but this, alas, I could not make him understand. So he treated me—his best friend—exactly as he treated the nest-robber and the bird-shooter.

I shall never know whether that nest contained eggs or young birds; or whether perchance there was no nest at all, and I had been deceived from the first by the most artful and beguiling of birds. And through all this I had never once squarely seen the chat I had been following.

> "Even yet thou art to me
> No bird, but, an invisible thing,
> A voice, a mystery."

XX. ON THE LAWN

The first thing that strikes an Eastern bird student in the Rocky Mountain region, as I have already said, is the absence of the birds he is familiar with. Instead of the chipping sparrow everywhere, one sees the lazuli-painted finch, or the Rocky Mountain bluebird; in place of the American robin's song, most common of sounds in country neighborhoods on the Atlantic side of the continent, is heard the silver bell of the towhee bunting, sometimes called marsh robin [Eastern towhee], or the harsh "chack" of Brewer's blackbird; the music that opens sleepy eyes at daybreak is not a chorus of robins and song-sparrows, but the ringing notes of the chewink, the clear-cut song of the Western meadow-lark, or the labored utterance of the black-headed grosbeak; it is not by the melancholy refrain of the whippoorwill or the heavenly hymns of thrushes that the approach of night is heralded, but by the cheery trill of the house wren or the dismal wail of the Western wood-pewee.

Most of all does the bird-lover miss the thrushes from the feathered orchestra. Some of them may dwell in that part of the world,—the books affirm it, and I cannot deny it,—but this I know: one whose eye is untiring, and whose ear is open night and day to bird-notes, may spend May, June, July, yes, and even August, in the haunts of Rocky Mountain birds, and not once see or hear either of our choice singing thrushes.

However the student may miss the birds he knows at home, he must rejoice in the absence of one,— the English sparrow. When one sees the charming purple finch and summer yellow-bird, nesting and singing in the streets of Denver, and the bewitching Arkansas goldfinch and the beautiful Western bluebird perfectly at home in Colorado Springs, he is reminded of what might be in the Eastern cities, if only the human race had not interfered with Nature's distribution of her feathered families. In Utah, indeed, we meet again the foreigner, for in that unfortunate Territory the man, wise in his own conceit, was found to introduce him, and Salt Lake, the city of their pride and glory, is as completely infested by the feathered tramp as New York itself. Happy is Colorado that great deserts form her borders, and that chains of mountains separate her from her neighbors; for, since the sparrow is as fond of the city as Dr. Johnson, it may be hoped that neither he, nor his children, nor his grandchildren, will ever cross the barriers.

In Utah, as everywhere, the English sparrows are sharp-witted rogues, and they have discovered and taken possession of the most comfortable place for bird quarters to be found, for protection from the terrible heat of summer, and the wind and snow of winter; it is between the roof and the stone or adobe walls of the houses. Wherever the inequalities of the stones or the shrinkage of the wood has left an opening, and made penetration possible, there an English sparrow has established a permanent abode.

The first bird I noticed in the quiet Mormon village where I settled myself to study was a little beauty in blue. I knew him instantly, for I had met him before in Colorado. He was dining luxuriously on the feathery seeds of a dandelion when I discovered him, and at no great distance was his olive-clad mate, similarly engaged. They were conversing cheerfully in low tones, and in a few minutes I suppose he called her attention to the superior quality of his dandelion; for she came to his side, and he at once flew to a neighboring bush and burst into song. It was a pretty little ditty, or rather a musical rattle on one note, resembling the song of the indigo bird, his near relative.

The lazuli-painted finch should be called the blue-headed finch, for the exquisite blueness of his whole head, including throat, breast, and shoulders, as if he had been dipped so far into blue dye, is his distinguishing feature. The bluebird wears heaven's color; so does the jay, and likewise the indigo bird; but not one can boast the lovely and indescribable shade, with its silvery reflections, that adorns the lazuli. Across the breast, under the blue, is a broad band of chestnut, like the breast color of our bluebird, and back of that is white, while the wings and tail are dark. Altogether, he is charming to look upon. Who would not prefer him about the yard to the squawking house sparrow, or even the squabbling chippy?

My catching the pair at dinner was not an accident; I soon found out that they lived there, and had settled upon a row of tall raspberry bushes that separated the garden from the lawn for their summer home. Madam was already at work collecting her building materials, and very soon the fragile walls of her pretty nest were formed in an upright crotch of the raspberries, about a foot below the top.

Naturally, I was greatly interested in the fairy house building, and often inspected the work while the little dame was out of sight. One day, however, as I was about to part the branches to look in, I heard an anxious "phit," and glanced up to see the owner alight on the lowest limb of a peach-tree near by. Of course I turned away at once, pretending that I was just passing, and had no suspicion of her precious secret in the raspberries, and hoping that she would not mind. But she did mind, very seriously; she continued to stand on that branch with an aggrieved air, as if life were no longer worth living, now that her home was perhaps discovered. Without uttering a sound or moving a muscle, so far as I could see, she remained for half an hour before she accepted my taking a distant seat and turning my attention to dragonflies as an apology, and ventured to visit her nest again. After that I made very sure that she was engaged elsewhere before I paid my daily call.

The dragonflies, by the way, were well worth looking at; indeed, they divided my interest with the birds. So many and such variety I never noticed elsewhere, and they acted exactly like fly-catching birds, staying an hour at a time on one perch, from which every

now and then they sallied out, sweeping the air and returning to the perch they had left. Sometimes I saw four
or five of them at once, resting on different dead twigs in
the yard the other side of the lawn, and I have even seen
one knock a fellow-dragonfly off a favorite perch and
take it himself.

They were very beautiful, too: some with wings of
transparent white or light amber barred off by wide
patches of rich dark brown or black; others, again, smaller, and all over blue as the lazuli's head; and a third of
brilliant silver, which sparkled as it flew, as if covered
with spangles. One alighted there with wings which
seemed to be covered with a close and intricate design in
the most brilliant gold thread. I went almost near
enough to put my hand on him, and I never saw a more
gorgeous creature; beside his beautiful wings his back
was of old gold, coming down in scallops over the black
and dark blue under part.

In due time four lovely blue eggs filled the nest of
the lazuli, and about the middle of June madam began to
sit, and I had to be more careful than ever in timing my
visits.

Some birds approach their nest in a loitering,
aimless sort of way, as if they had no particular business,
in that quarter, and, if they see any cause for alarm, depart with an indifferent air that reveals nothing of their
secret. Not thus the ingenuous lazuli. She showed her
anxiety every moment; coming in the most businesslike
way, and proclaiming her errand to the most careless observer, till I thought every boy on the street would know
where her eggs were to be found. She had a very pretty
way of going to the nest; indeed, all her manners were

winning. She always alighted on the peach-tree branch, looked about on all sides, especially at me in my seat on the piazza, flirted her tail, uttered an anxious "phit," and then jumped off the limb and dived under the bushes near the ground. It is to be presumed that she ascended to her nest behind the leaves by hopping from twig to twig, though this I could never manage to see.

And what of her gay little spouse all this time? Did he spend his days cheering her with music, as all the fathers of feathered families are fabled to do? Indeed he did not, and until I watched very closely, and saw him going about over the poplars in silence, I thought he had left the neighborhood. Once in the day he had a good singing time, about five o'clock in the morning, two hours before the sun rose over the mountains. If one happened to be awake then, he would hear the most rapturous song, delivered at the top of his voice, and continuing for a long time. But as it grew lighter, and the human world began to stir, he became quiet again, and, if he sang at all, he went so far from home that I did not hear him.

But the wise little blue-head had not deserted; he was merely cautious. Every time that the little sitter went off for food she met him somewhere, and he came back with her. Occasionally he took a peep at the treasures himself, but he never entered by her roundabout way. He always flew directly in from above.

Ten days passed away in this quiet manner, my attention divided between the birds, the dragonflies, and the clacking grasshopper, who went jerking himself about with a noise like a subdued lawn-mower, giving one the impression that his machinery was out of order.

The tenth day of sitting we had a south wind. That does not seem very terrible, but a south wind on the shore of the Great Salt Lake is something to be dreaded.

"A wind that is dizzy with whirling play,
A dozen winds that have lost their way."

It starts up suddenly, and comes with such force as to snap off the leaves of trees, and even the tender twigs of shrubs. As it waxes powerful it bends great trees, and tries the strength of roofs and chimneys. From the first breath it rolls up tremendous clouds of dust, that come and come, and never cease, long after it seems as if every particle in that rainless land must have been driven by. It is in the "Great Basin," and the south wind is the broom that sweeps it clean. Not only dust does the south wind bring, but heat, terrible and suffocating, like that of a fiery furnace. Before it the human and the vegetable worlds shrink and wither, and birds and beasts are little seen.

Such a day was the birthday in the little nest in the raspberries, and on my usual morning call I found four featherless birdlings, with beaks already yawning for food. Every morning, of course, I looked at the babies, but it was not till the eighth day of their life that I found their eyes open. Before this they opened their mouths when I jarred the nest in parting the branches, thus showing they were not asleep, but did not open their eyes, and I was forced to conclude that they were not yet unclosed.

Sometimes the daily visit was made under difficulties, and I was unpleasantly surprised when I stepped upon the grass of the little lawn that I was obliged to cross. The grass looked as usual; the evening before we had been sitting upon it. But all night a stream had been silently spreading itself upon it, and my hasty step was into water two or three inches deep, which swished up in a small fountain and filled a low shoe in an instant.

This is one of the idiosyncrasies of irrigation, which it seemed I should never get accustomed to, and several times I was obliged to turn back for overshoes before I could pay my usual call. A lawn asoak is a curious sight, and always reminds me of Lanier's verses,

> "A thousand rivulets run
> 'Twixt the roots of the soil;
> the blades of the marsh grass stir;
> ... and the currents cease to run,
> And the sea and the marsh are one."

The morning the lazulis were ten days old, before I came out of the house, that happened which so often puts an end to a study of bird life,—the nest was torn out of place and destroyed, and the little family had disappeared. The particulars will never be known. Whether a nest-robbing boy or a hungry cat was the transgressor, and whether the nestlings were carried off or eaten, or had happily escaped, who can tell? I could only judge by the conduct of the birds themselves, and as they did not appear disturbed, and continued to carry food, it is to be presumed that part, if not all, of the brood was saved from the wreck of their home.

Happily, to console me in my sorrow for this catastrophe, the lazuli was not the only bird to be seen on the lawn, though his was the only nest. I had for some time been greatly interested in the daily visits of a humming-bird, a little dame in green and white, who had taken possession of a honeysuckle vine beside the door, claiming the whole as her own, and driving away, with squeaky but fierce cries, any other of her race who ventured to sip from the coral cups so profusely offered.

The season for humming-birds opened with the locust blossoms next door, which were for days a mass of blooms and buzzings, of birds and bees. But when the fragrant flowers began to fall and the ground was white with them, one bird settled herself on our honeysuckle, and there took her daily meals for a month. Being not six feet from where I sat for hours every day, I had the first good opportunity of my life to learn the ways of one of these queer little creatures in feathers.

After long searching and much overhauling of the books, I made her out to be the female broad-tailed humming-bird, who is somewhat larger than the familiar ruby-throat of the East. Her mate, if she had one, never came to the vine; but whether she drove him away and discouraged him, or whether he had an independent source of supply, I never knew. She was the only one whose acquaintance I made, and in a month's watching I came to know her pretty well.

In one way she differed strikingly from any humming-bird I have seen: she alighted, and rested frequently and for long periods. Droll enough it looked to see such an atom, such a mere pinch of feathers, conduct herself after the fashion of a big bird; to see her wipe that needle-like beak, and dress those infinitesimal feathers, combing out her head plumage with her minute black claws, running the same useful appendages through her long, gauzy-looking wings, and carefully removing the yellow pollen of the honeysuckle blooms which stuck to her face and throat. Her favorite perch was a tiny dead twig on the lowest branch of a poplar-tree, near the honeysuckle. There she spent a long time each day, sitting usually, though sometimes she stood on her little wiry legs.

But though my humming friend might sit down, there was no repose about her; she was continually in motion. Her head turned from side to side, as regularly, and apparently as mechanically, as an elephant weaves his great head and trunk. Sometimes she turned her attention to me, and leaned far over, with her large, dark eyes fixed upon me with interest or curiosity. But never was there the least fear in her bearing; she evidently considered herself mistress of the place, and reproved me if I made the slightest movement, or spoke too much to a neighbor. If she happened to be engaged among her honey-pots when a movement was made, she instantly jerked herself back a foot or more from the vine, and stood upon nothing, as it were, motionless, except the wings, while she looked into the cause of the disturbance, and often expressed her disapproval of our behavior in squeaky cries.

The toilet of this lilliputian in feathers, performed on her chosen twig as it often was, interested me greatly. As carefully as though she were a foot or two, instead of an inch or two long, did she clean and put in order every plume on her little body, and the work of polishing her beak was the great performance of the day. This member was plainly her pride and her joy; every part of it, down to the very tip, was scraped and rubbed by her claws, with the leg thrown over the wing, exactly as big birds do. It was astonishing to see what she could do with her leg. I have even seen her pause in mid-air and thrust one over her vibrating wing to scratch her head.

Then when the pretty creature was all in beautiful order, her emerald-green back and white breast immaculate, when she had shaken herself out, and darted out and drawn back many times her long bristle-like tongue, she would sometimes hover along before the tips of the fence-stakes, which were like laths, held an inch apart by wires,—collecting, I suppose, the tiny spiders which were to be found there. She always returned to the honeysuckle, however, to finish her repast, opening and closing her tail as one flirts a fan, while the breeze made by her wings agitated the leaves for two feet around her. Should a blossom just ready to fall come off on her beak like a coral case, as it sometimes did, she was indignant indeed; she jerked herself back and flung it off with an air that was comical to see.

When the hot wind blew, the little creature seemed to feel the discomfort that bigger ones did: she sat with open beak as though panting for breath; she flew around with legs hanging, and even alighted on a convenient leaf or cluster of flowers, while she rifled a blossom, standing with sturdy little legs far apart, while stretching up to reach the bloom she desired.

Two statements of the books were not true in the case of this bird: she did not sit on a twig upright like an owl or a hawk, but held her body exactly as does a robin or sparrow; and she did fly backward and sideways, as well as forward.

Toward the end of June my tiny visitor began to make longer intervals between her calls, and when she did appear she was always in too great haste to stop; she passed rapidly over half a dozen blossoms, and then flitted away. Past were the days of loitering about on poplar twigs or preening herself on the peach-tree. It was plain that she had set up a home for herself, and the mussy state of her once nicely kept breast feathers told the tale, —she had a nest somewhere. Vainly, however, did I try to track her home: she either took her way like an arrow across the garden to a row of very tall locusts, where a hundred humming-birds' nests might have been hidden, or turned the other way over a neighbor's field to a cluster of thickly grown apple-trees, equally impossible to search. If she had always gone one way I might have tried to follow, but to look for her infinitesimal nest at opposite poles of the earth was too discouraging, even if the weather had been cool enough for such exertion.

When at last I could endure the wind and the dust and the heat no longer, and stood one morning on the porch, waiting for the most deliberate of drivers with his carriage to drive me to the station, that I might leave Utah altogether, the humming-bird appeared on the scene, took a sip or two out of her red cups, flirted her feathers saucily in my very face, then darted over the top of the cottage and disappeared; and that was the very last glimpse I had of the little dame in green.

INDEX

Idle
Winter
Press

www.ingramcontent.com/pod-product-compliance
Lightning Source LLC
Chambersburg PA
CBHW061018280326
41935CB00009B/1011